T0192101

Lecture Notes
in Business Information Processing   **408**

More information about this series at http://www.springer.com/series/7911

Adam Przybyłek · Jakub Miler ·
Alexander Poth · Andreas Riel (Eds.)

Lean and Agile Software Development

5th International Conference, LASD 2021
Virtual Event, January 23, 2021
Proceedings

 Springer

Editors
Adam Przybyłek 🆔
Gdańsk University of Technology
Gdańsk, Poland

Jakub Miler 🆔
Gdańsk University of Technology
Gdańsk, Poland

Alexander Poth 🆔
Volkswagen AG
Wolfsburg, Germany

Andreas Riel 🆔
Université Grenoble Alpes
Grenoble, France

ISSN 1865-1348   ISSN 1865-1356 (electronic)
Lecture Notes in Business Information Processing
ISBN 978-3-030-67083-2   ISBN 978-3-030-67084-9 (eBook)
https://doi.org/10.1007/978-3-030-67084-9

This Springer imprint is published by the registered company Springer Nature Switzerland AG
The registered company address is: Gewerbestrasse 11, 6330 Cham, Switzerland

Preface

As COVID-19 swept the world in early 2020, countries began putting their citizens under partial or total lockdown, while software companies started implementing remote-working policies for many or all of their employees. The pandemic had an almost immediate effect on agile software development, since agile work practices are harder to perform when casual conversations are limited due to the online nature of meetings and interactions. Nevertheless, more than ever before, the agile mindset and practices have turned out to be vital for organizations to navigate through the new terrain. In this setting, the LASD (Lean and Agile Software Development) conference series and the community feel particularly proud of their contributions to research and practice investigating how to stay Agile while working remotely during the pandemic.

In 2021, LASD separated from FedCSIS and became a standalone conference that took place virtually on 23 January 2021 as an online event. As everyone involved in LASD 2021 worked voluntarily, the conference was fully free of charge. LASD 2021 received 32 submissions. After a rigorous review process, which included at least 3 reviews per submission, 10 high-quality full papers and 2 short papers were selected. The accepted papers were presented to a well-focused audience, thus the discussion provided the authors with new ideas and directions for further research. Topics discussed in this volume range from teams under COVID-19 through women in Agile, to product roadmapping and non-functional requirements.

Corresponding authors of all accepted papers received a complimentary 1 year membership in Agile Alliance. Agile Alliance is a nonprofit global member organization dedicated to promoting the concepts of Agile software development as outlined in the Agile Manifesto. With more than 75,000 members and subscribers around the globe, Agile Alliance is driven by the principles of Agile methodologies and the value delivered to developers, business, and end users. Agile Alliance organizes and supports events to bring the Agile community together on an international scale. Besides, the corresponding author of the best paper received a prize of a TeamRetro single team access for 12 months.

The high quality of the LASD 2021 technical program was enhanced by two keynote lectures delivered by outstanding guests: Philipp Diebold ("Agility Yesterday, Today & Tomorrow") and Sanjay Misra ("Pair Programming: An Empirical investigation in an Agile Software Development environment").

To live the agile mindset, the LASD conference focuses on highly relevant research outcomes and fosters their way into practice. We believe the highest value produced by a conference is that researchers' outcomes are pulled by practitioners from industry to integrate them into innovative products and services.

We would like to express our gratitude to everyone who made LASD 2021 successful. First of all, we thank all authors for their contributions, the members of the Program Committees for taking the time and effort to provide insightful remarks, as well as both keynote speakers for their impressive speeches. We are also deeply grateful to Ivan Luković for the opportunity to publish an extended version of the best paper in Computer Science and Information Systems (ComSIS). Furthermore, we acknowledge Dominik Grzegorzek for LaTeX typesetting assistance. Finally, we would like to thank the team at Springer (especially Ralf Gerstner, Christine Reiss, Alfred Hofmann, Anna Kramer, Guido Zosimo-Landolfo, Ramvijay Subramani, and Anja Seibold) for making this volume possible.

We hope that you find this monograph useful for your professional and academic activities, and we wish you a stimulating read. We also cordially invite you to visit our conference website at https://lasd.pl, and to join us for the upcoming edition.

January 2021

Adam Przybyłek
Jakub Miler
Alexander Poth
Andreas Riel

Organization

Conference and Program Committee Chair

Adam Przybyłek   Gdańsk University of Technology, Poland

Program Committee

Muhammad Ovais Ahmad	Karlstad University, Sweden
Ibrahim Akman	Atılım University, Turkey
Sikandar Ali	China University of Petroleum, China
Fernando Almeida	University of Porto & INESC TEC, Portugal
Mohammad Alshayeb	King Fahd University of Petroleum and Minerals, Saudi Arabia
Samuil Angelov	Fontys University of Applied Sciences, The Netherlands
Irena Bach-Dabrowska	WSB Gdańsk, Poland
Alessandra Bagnato	SOFTEAM R&D Department, France
Woubshet Behutiye	University of Oulu, Finland
Alvine Boaye Belle	École de Technologie Supérieure, Canada
Nourchene Elleuch Ben Ayed	Higher Colleges of Technology, UAE
Mario Bernhart	Vienna University of Technology, Austria
Vikram Bhadauria	Texas A&M University Texarkana, USA
Nik Nailah Binti Abdullah	Monash University Malaysia, Malaysia
Miklós Biró	Software Competence Center Hagenberg and Johannes Kepler University Linz, Austria
Jan Olaf Blech	Aalto University, Finland
Markus Borg	SICS Swedish ICT AB, Sweden
Alena Buchalcevova	Prague University of Economics and Business, Czech Republic
Jim Buchan	Auckland University of Technology, New Zealand
Luigi Buglione	Engineering Ingegneria Informatica SpA, Italy
Daniela Cruzes	SINTEF ICT, Norway
Wiktor Bohdan Daszczuk	Warsaw University of Technology, Poland
Igor Dejanović	Faculty of Technical Sciences, Serbia
Anna Derezinska	Warsaw University of Technology, Poland
Philipp Diebold	Bagilstein GmbH, Germany
Arpita Dutta	IIT Kharagpur, India
Maria Jose Escalona	Universidad de Sevilla, Spain
Imane Essebaa	Hassan II University of Casablanca, Morocco
Fabian Fagerholm	Aalto University, Finland

Fernando Marques Figueira Filho	Universidade Federal do Rio Grande do Norte, Brazil
Gabriel Alberto García-Mireles	Universidad de Sonora, Mexico
Javad Ghofrani	University of Applied Sciences Dresden, Germany
Krzysztof Goczyła	Gdańsk University of Technology, Poland
Sangharatn Godboley	NIT Rourkela, India
Javier Gonzalez Huerta	Blekinge Institute of Technology, Sweden
Peggy Gregory	University of Central Lancashire, UK
Janusz Górski	Gdańsk University of Technology, Poland
Ridewaan Hanslo	Council for Scientific and Industrial Research, South Africa
Sebastian Heil	Chemnitz University of Technology, Germany
Andreas Hinderks	University of Seville, Spain
Uwe Hohenstein	Siemens AG, Germany
Philipp Hohl	ZF Friedrichshafen AG, Germany
Marko Ikonen	Projektivarikko Oy, Finland
Irum Inayat	National University of Computer and Emerging Sciences, Pakistan
Andrea Janes	Free University of Bozen-Bolzano, Italy
Aleksander Jarzębowicz	Gdańsk University of Technology, Poland
Miloš Jovanović	University of Novi Sad, Serbia
Janne Järvinen	F-Secure Corporation, Finland
George Kakarontzas	Aristotle University of Thessaloniki, Greece
Kalinka Kaloyanova	Sofia University, Bulgaria
Benjamin Kanagwa	Makerere University, Uganda
Georgia Kapitsaki	University of Cyprus, Cyprus
Matěj Karolyi	Masaryk University, Czech Republic
Aleksandra Karpus	Gdańsk University of Technology, Poland
Mohamad Kassab	Innopolis University, Russia
Marija Katić	Birkbeck, University of London, UK
Wiem Khlif	University of Sfax, Tunisia
Sylwia Kopczyńska	Poznań University of Technology, Poland
Martin Kropp	University of Applied Sciences and Arts Northwestern Switzerland, Switzerland
Pasi Kuvaja	University of Oulu, Finland
Maarit Laanti	Nitor, Finland
Timo O. A. Lehtinen	Aalto University, Finland
Valentina Lenarduzzi	LUT University, Finland
Grischa Liebel	Reykjavik University, Iceland
Ivan Luković	University of Novi Sad, Serbia
Ilaria Lunesu	Università degli Studi di Cagliari, Italy
Katarzyna Łukasiewicz	Gdańsk University of Technology, Poland
Viljan Mahnič	University of Ljubljana, Slovenia
George Mangalaraj	Western Illinois University, USA
Bartosz Marcinkowski	University of Gdańsk, Poland

Christoph Matthies	Hasso Plattner Institute at the University of Potsdam, Germany
Manuel Mazzara	Innopolis University, Russia
Antoni-Lluís Mesquida Calafat	University of the Balearic Islands, Spain
Jakub Miler	Gdańsk University of Technology, Poland, Poland
Gloria Miller	Skema Business School, France
Sanjay Misra	Covenant University, Nigeria
Durga Prasad Mohapatra	NIT Rourkela, India
Miguel Ehecatl Morales Trujillo	University of Canterbury, New Zealand
Richard Mordinyi	Vienna University of Technology, Austria
Karolina Muszyńska	University of Szczecin, Poland
Mirna Muñoz	Centro de Investigación en Matemáticas, Mexico
Jürgen Münch	Reutlingen University, Germany
Yen Ying Ng	Nicolaus Copernicus University, Poland
Anh Nguyen-Duc	University of South-Eastern Norway, Norway
Arne Noyer	Osnabrück University and Willert Software Tools GmbH, Germany
Hanna Oktaba	National Autonomous University of Mexico, Mexico
Marco Ortu	University of Cagliari, Italy
Tosin Daniel Oyetoyan	Western Norway University of Applied Sciences, Norway
Necmettin Özkan	Kuveyt Türk Participation Bank, Turkey
Subhrakanta Panda	Birla Institute of Technology and Science, Pilani, India
Rui Humberto R. Pereira	Instituto Politécnico do Porto - ISCAP, Portugal
Kesava Pillai	Asia Pacific University of Technology and Innovation, Malaysia
Aneta Poniszewska-Maranda	Lodz University of Technology, Poland
Alexander Poth	Volkswagen AG, Germany
Michał Przybyłek	University of Warsaw, Poland
Raman Ramsin	Sharif University of Technology, Iran
Andreas Riel	Grenoble Alpes University, France
Sonja Ristić	University of Novi Sad, Serbia
Bruno Rossi	Masaryk University, Czech Republic
Zdenek Rybola	FIT CTU in Prague, Czech Republic
Dina Salah	Sadat Academy for Management Sciences, Egypt
Mattia Salnitri	University of Trento, Italy
Wylliams Barbosa Santos	University of Pernambuco, Brazil
Eva-Maria Schön	University of Seville, Spain
Jorge Sedeno	University of Seville, Spain
Mali Senapathi	Auckland University of Technology, New Zealand

Illia Shkroba	Polish-Japanese Academy of Information Technology, Poland
Marcin Sikorski	Polish-Japanese Academy of Information Technology, Poland
Michel Soares	Federal University of Sergipe, Brazil
Álvaro Soria	ISISTAN Research Institute, Argentina
Maria Spichkova	RMIT University, Australia
Olga Springer	Gdańsk University of Technology, Poland
Christoph Johann Stettina	Leiden University, The Netherlands
Tor Stålhane	Norwegian University of Science and Technology, Norway
Julian Szymański	Gdańsk University of Technology, Poland
Michał Śmiałek	Politechnika Warszawska, Poland
Davide Taibi	Free University of Bozen-Bolzano, Italy
Ayca Tarhan	Hacettepe University, Turkey
Adel Taweel	Birzeit University, Palestine
Sven Theobald	Fraunhofer IESE, Germany
Jörg Thomaschewski	University of Applied Sciences Emden/Leer, Germany
Carlos Torrecilla Salinas	University of Seville, Spain
Michael Unterkalmsteiner	Blekinge Institute of Technology, Sweden
Andrzej Wardziński	Gdańsk University of Technology, Poland
Paweł Weichbroth	Gdańsk University of Technology, Poland
Jan Werewka	AGH University of Science and Technology, Poland
Dominique Winter	University of Applied Sciences Emden/Leer, Germany
Michał Wróbel	Gdańsk University of Technology, Poland
Włodzimierz Wysocki	West Pomeranian University of Technology, Poland
Murat Yilmaz	Çankaya University, Turkey
Nacer Eddine Zarour	Constantine 2 University, Algeria

Additional Reviewers

Jannik Fangmann	University of Applied Sciences Emden/Leer, Germany
Hanna Looks	University of Applied Sciences Emden/Leer, Germany
Paweł Markowski	Polish-Japanese Academy of Information Technology, Poland

Contents

Short Papers

Keynote Paper

Full Papers

Women in Agile: The Impact of Organizational Support for Women's Advancement on Teamwork Quality and Performance in Agile Software Development Teams

Asli Yüksel Aksekili[2(✉)] and Christoph Johann Stettina[1,2]

[1] Leiden Institute of Advanced Computer Science, Leiden University,
Niels Bohrweg 1, 2333 CA Leiden, The Netherlands
[2] Accenture—SolutionsIQ Netherlands,
Orteliuslaan 1000, 3528 BD Utrecht, The Netherlands
yukselaksekili@gmail.com

Abstract. In this research we investigate how organizational support for women's advancement and gender diversity as a headcount is related to teamwork quality and team performance based upon Hoegl and Gemuenden's teamwork quality model (TWQ). Using an online survey we obtained data from 77 professionals working in agile software development teams. The results show that organizational support for women's advancement has a positive impact on all TWQ dimensions, while gender diversity expressed as an equal headcount number has only a positive impact on coordination and balance of member contributions in our data set. Further, in line with previous research, our data shows that all dimensions of the TWQ model have a positive impact on team performance in agile software development teams. Our findings indicate that an organization's mindset towards gender diversity has a stronger effect on team performance than gender diversity as a headcount number only.

Keywords: Gender diversity · Gender equality · Teamwork quality · Team performance · Agile software development

1 Introduction

Practically every one of us has had to deal with the gender equality question in teams in one way or another. While several studies point out the impact of gender headcount on teamwork in general as well as in software development in particular, the support of an organization for women's advancement is a critical aspect of gender equality that is far less understood [1].

Understanding the impact of an organization's mindset towards supporting women's career advancement rather than looking only at gender headcount number as a snapshot in time is interesting in the context of agile teams. Agile

© Springer Nature Switzerland AG 2021
A. Przybyłek et al. (Eds.): LASD 2021, LNBIP 408, pp. 3–23, 2021.
https://doi.org/10.1007/978-3-030-67084-9_1

frameworks put a strong emphasis on the social nature of work and the impact of human factors on team performance, and rely heavily on self-management, communication, and coordination [2–4]. Evidence suggests that gender diversity in software teams is positive and significant predictors of productivity [5], but the presence of women in software teams generally also reduces the amount of community smells like Organizational Silo effects (overly disconnected subgroups) or Lone Wolves (defiant community members) [6]. These effects are interesting for the long-term performance of agile teams.

In this paper, we examine the impact of gender diversity as a headcount and the organizational support for women's advancement on the teamwork qualities of agile software development teams. Hoegl and Gemuenden's [7] teamwork quality and performance model provides relevant variables for the purpose of this study and with its suitability to agile software development teams [8]. Based on responses of 77 agile software development professionals, our findings suggest that (1) gender diversity has a significant impact on the coordination and balance of member contributions in agile teams, and (2) organizational support for women's advancement has a significant effect on all variables.

In the remainder of the paper, we review the related work, explain the variables of the selected model in detail and the hypotheses on the conceptual model, outline the methodology and data collection, and demonstrate the descriptive statistics and the statistical analysis of the results. Finally, we discuss the results and give final comments on the direction of research.

2   Related Work

In this section, we discuss the social nature of agile teams, provide an overview of previous research on women's skills in relation to software development and agile methodologies, and elaborate the research gap.

2.1   Gender Diversity in Software Teams

The relation of women to studies and to work in general as well as to agile software development in particular has been studied by numerous researchers [9–15]. Various aspects such as differences in the work and leadership styles of men and women [12] or the impact of gender diversity on the ongoing project results have been analyzed qualitatively and quantitatively.

While many organizations introduce gender equality programs including targets for a gender balanced workforce (e.g. a workforce that is equally 50 percent women and 50 percent men, or specific targets for senior management/leadership roles) [16,17], available studies suggest that the effect of gender diversity on performance varies significantly across countries and industries due to differences in institutional contexts [13]. In some contexts, for example, studies indicate that teams with a rather unbalanced gender composition outperformed teams with a balanced (50/50) or all-male and all-female teams [14,15].

Although some studies have concluded that men and women show more similarities than disparities, differences have been shown by empirical research do exist due to biological, neurological, and psychological dimensions [18]. For example, a meta-analysis conducted by Eagly and Johnson [19] of more than 160 studies of sex-related differences found that women in the workplace engage in a more collaborative and democratic and less commanding and directive style than men do. This tendency does, however, decline in highly male-dominated environments. Beranek et al. [20] found that in software development teams, a female management style centered on the communication and creation of team interactions, more often assuming a team-building and maintenance role than males in the same positions.

Agile methodologies require software development teams to possess an advanced level of teamwork qualities to perform better, in addition to an aptitude in testing, short releases, customer satisfaction, and a sustainable work-pace [21]. Considering that recent research studies defined women's style of management as typically collaborative and communicative, agile software development can therefore benefit from the gender diversity in this respect. A study conducted by the Israel Institute of Technology [22] showed how the agile approach towards software development creates an environment in which women and men communicate similarly and therefore support gender diversity. Many organizations relying on virtual collaboration apply agile methods [23].

Weilemann and Brune [24] investigated the behavior of female Scrum Masters with an exploratory qualitative study of a student's software project. They found that female Scrum Masters understood the real needs of the team members, respected them, and made accurate judgments while including everyone in the decision-making process.

2.2 Organizational Support for Women's Advancement

Involving more women does not automatically generate a better team; the expertise of its members must of course be considered. But if females see that they are rewarded for what they do, they are more likely to find the incentive to align their expertise with their role responsibilities.

According to Ruderman and Ohlott [25], organizational environments are themselves gendered, adjusting themselves to accommodate and develop their male members, while females' contribution, performance, and success are measured and evaluated differently [18]. This trend continues despite the preponderance of studies suggesting that the most successful organizations are the ones that value the contributions made by females and provide development opportunities for their talents [26]. Moreover, Jawahar and Hemmasi [1] introduced an empirical work with their study among members of US-based national women's associations on the impact on turnover intentions of the perceived organizational support for women's advancement. In this study, based on the social exchange theory of Blau [27], they claimed that the organizations might be sacrificing their female talents by failing to support career advancements. According to this theory, the relationship with the organization is of great significance for the motivation of the employee. Jawahar and Hemmasi found that if women do not perceive the support of the organization, they might be less likely to remain with it. In addition, they observed a positive and significant relationship between the organization's support for women's advancement and women's job satisfaction and employer satisfaction respectively.

Previous studies had underlined the absence of available career paths drawn for females in organizations [10]. To attract and retain female professionals, e.g. organizations must support diversity in their promotion policies. Thus, their suggestion was that software development teams should be built with consideration for gender roles. Moreover, they underlined that in general, the companies' understanding of gender diversity is to hire more women IT professionals without paying attention to specific gender roles.

2.3   Gaps in the Literature and Research Objectives

Gender equality in organizations has been a commonly addressed topic in companies' agendas in the past few decades and every organization's approach to this issue has differed. However, the general focus has been on closing the headcount gap, which is the most obvious change to implement after several years of male domination. Razavian and Lago [10] placed an emphasis on the less visible yet more harmful aspects of gender inequality, which perhaps can also be defined as the next step in the issue. Studies indicate that inequality also stems from organizations' failure to support women's advancement that in turn leads to an unwillingness to perform to their best, job and employer dissatisfaction [1] and ultimately to a loss of talent. This is a disadvantage for the organization itself, but is a direct result of the prevailing culture of the organization itself.

Despite a few studies exploring the impact of gender on agile software development teams, published works on this relationship is limited. Hence, we pose the following two research questions: *(1) Are gender diversity as a headcount and the organizational support for women's advancement related to communication, coordination, balance of member contributions, mutual support, effort, cohesion, and agility in agile software development teams? (2) How are these teamwork qualities individually related to the team's performance in such environments?*

3   Conceptual Model

In this section, we propose our conceptual model associated with our research questions and then describe the selected variables used with its operationalization based on previous studies. For each variable, we first define it and explain its relation with both agile software development and women's capabilities before putting forward the hypotheses.

This study proposes a research model in combination with Hoegl and Gemuenden's [7] team performance model to investigate both the impact of gender diversity as a headcount and the organization's support for the advancement of women individually on teamwork quality variables and team agility which constituted the Phase I of the study. In Phase II, the impact of these teamwork quality and agility variables on the team performance is also examined. In total, 21 hypotheses were tested. Figure 1 presents a visual representation of the model.

Gender Diversity as a Headcount. Does gender diversity matter for team process? While this question has been studied previously (cf. [5,28,29]), the common finding was that the social orientation of women diminishes team conflicts by reducing egocentric listening and allowing for a higher quality of teamwork [14] with equal participation of men and women [30–32]. Thus, the socio-emotional behavior and non-aggressive strategies associated with women in teams can lead to an improved teamwork and agility as the gender headcount approaches equal numbers in agile software development teams. We test this impact of gender diversity in this study.

Organizational Support for Women's Advancement. Jawahar and Hemmasi [1] described perceived organizational support as the degree to which employees perceive that the organization values their work, respects them and is concerned for their well being. They studied its impact in terms of women's advancement on the females' intentions to leave the organization. They found a negative relationship and observed that an organization's support had a positive impact on the women's degree of job and employer satisfaction. Since several earlier studies had found positive correlation between performance and job satisfaction [33], it seems reasonable to claim a similar relationship for the teamwork qualities of agile teams, and organizations' support for women's advancement.

Communication. Communication has been discussed as a factor of teams' project success within several studies [34,35] as well as in agile software development [36,37]. In 2015, Razavian and Lago's [10] study revealed that in software architecture teams, the quality of communication is seen as a result of the feminine expertise of eliciting the real needs of the team and questioning the problems and constraints. Therefore, we expect both gender diversity as a headcount and the organization's support for women's advancement to have a positive impact on the quality of communication in agile teams. *Hypothesis 1:* Gender diversity as a headcount has a positive impact on the quality of communication. *Hypothesis 2:* Organizational support for women's advancement is positively related to the quality of communication.

Coordination. Coordination, one important component of the quality of teamwork, can be defined as a common understanding of the interrelatedness and current status of members' contributions within teams [7]. In agile teams, tasks are often selected and delegated when planning a new iteration which are then assigned to members who are expected to execute them in coordination with each other [8]. In 2015, research by Weilemann and Brune [24] showed that when females are assigned to Scrum master roles within software development teams, they demonstrate superior coordinating skills than male Scrum masters. Hence, by assuming that females are capable of improving the coordination within teams we argued that agile teams' coordination skills can be positively affected by gender diversity as a headcount and the organization's support for women's advancement can also have a positive impact on coordination. *Hypothesis 3:* Gender diversity as a headcount has a positive impact on the quality of coordination. *Hypothesis 4:* Organizational support for women's advancement is positively related with the quality of a team's coordination.

Balance of Member Contributions. It is an advantage for teams to know what task-related knowledge and experience are possessed by members to contribute to the decision-making process [38]. For agile software development teams with members who have different expertise (core development, system architecture, testing, etc.), balanced contribution throughout project works is critical [8]. The work of Rogelberg and Rumery [14] found that gender diverse teams have a superior quality of decision-making especially with the contribution of socioemotional behavior of females. Therefore, for this study, the expectation is that agile teams have a better balance of member contributions as gender diversity as a headcount increases and also as organizations' support for the advancement of women increases. *Hypothesis 5:* Gender diversity as a headcount has a positive impact on the balance of member contributions. *Hypothesis 6:* Organizational support for women's advancement is positively related to the balance of member contributions.

Mutual Support. The intensive collaboration of individuals in agile teams is dependent on cooperation rather than competition. Tjosvold [39] argued that for interdependent tasks, mutual support is very important for productivity. Thus, displaying mutual respect, granting assistance when needed, and further developing the ideas of other members through discussion rather than trying to outdo each other should be expected for a high-level team performance. The study of Eagly and Johannesen-Schmidt [12] argued that the behavior of females is more interpersonally oriented and democratic while in contrast, the behavior of males may be more autocratic and competitive. Thus, the equal presence of females can increase the mutual support in agile teams. Moreover, we also argue that as an organizations' support increases for the advancement of women, the mutual support in an agile team can increase. *Hypothesis 7:* Gender diversity as a headcount has a positive impact on mutual support within the team. *Hypothesis 8:* Organizational support for women's advancement is positively related to mutual support within the team.

Effort. Knowing that all team members are doing their best to support the tasks of the team is an aspect of teamwork quality [38]. Sharing the workload and prioritizing the team's work over other obligations are indicators of team members' commitments to common tasks. Within an organization where the employees perceive that everyone is being equally supported in their advancement, it is reasonable to expect a more open interaction and shared commitment that can create the conditions for a mutually supportive effort. Moreover, previous research indicates that teams, with equal participation and performance of males and females, lead to the best outcome. Therefore, we expect gender diversity to have a positive impact on effort in agile teams. *Hypothesis 9:* Gender diversity as a headcount has a positive impact on team effort. *Hypothesis 10:* Organizational support for women's advancement is positively related to team effort.

Cohesion. Mudrack [40] defined team cohesion as a dynamic process that involves the tendency of the group to stick together and remain united towards common goals and objectives. It is also referred to by Cartwright [35] as the degree to which team members desire to stay with the team. Within an agile team context, constant feedback mechanism is one of the factors that contributes to team awareness and commitment to the team goal which consequently leads to a cohesive team. Previous research suggests that females are more socially oriented in teams [41]. An enhanced level of interpersonal attraction among team members, and therefore a more cohesive environment can be expected when a gender diverse team is present. Moreover, considering the negative impact of organizational support for women's advancement on intentions of females to leave the organization [1] and also its positive impact on job and employer satisfaction, it is reasonable to expect its positive impact on cohesiveness in agile teams. *Hypothesis 11:* Gender diversity as a headcount has a positive impact on cohesion within the team. *Hypothesis 12:* Organizational support for women's advancement is positively related to cohesion within the team.

Team Agility. Team agility can be defined as the aptitude of a team to respond to rapidly changing business conditions and achieve successful exploration of competitive bases in a timely manner (e.g. speed, flexibility) [42] and empowerment is especially important for team agility [43]. Within a team context, members with stronger empowerment can become cognitively more flexible and more likely to respond to urgent issues with concrete and creative solutions, hence contribute into the team agility [44]. Based on these considerations, providing an equal work environment in terms of equal hiring and equal support for advancement, the agility of teams can be significantly higher. *Hypothesis 13:* Gender diversity as a headcount has a positive impact on team agility. *Hypothesis 14:* Organizational support for women's advancement is positively related to team agility.

Team Performance. Hoegl and Gemuenden [7] defined team performance as a team's ability to meet its established quality, cost, and time objectives and showed the positive impact of the quality of teamwork on it. In their 2016 study, Lindsjørn et al. [8] adapted the same model for agile teams and observed a similar impact of teamwork quality on performance. Hence, they reported that the model scales are suitable for both traditional and agile software development. In this study, we examine the impact of each teamwork quality subconstructs and the team agility on a team's performance and expect positive relationships. *Hypothesis 15, 16, 17, 18, 19, 20, 21:* The quality of communication *(H15)*, coordination *(H16)*, balance of member contributions *(H17)*, mutual support *(H18)*, effort *(H19)*, cohesion *(H20)*, and team agility *(H21)* are positively related to team performance.

4   Method

The conceptual model is tested with survey data collected from professionals who are working in agile software development teams. The reason for pursuing this methodology is that the surveys are reliable tools to collect real-world data and easy to execute. No limitation is set in terms of gender, years of experience, geography, or company. The online tool Qualtrics was used to write and distribute the survey through an anonymous link.

Participants. The target respondents of the survey were the practitioners working in agile software development teams with roles such as Software Developer, Scrum Master, Product Owner, etc. In total, 164 responses were received of which 77 were completed and 87 were incomplete. Among these 77 complete surveys, 33 respondents preferred to state their organization's name. Those 33 respondents named 17 different companies with the majority coming from the following sectors: Technology (57%), Consulting (24%), and Financial Services (12%). Five respondents preferred not to answer the demographics questions but since they had completed the first section, we included their data. Key characteristics of the participant data is shown on Table 1.

Table 1. Participant demographics

Roles	N	Gender (%)			Experience (Mean, in years)	
		M	F	Other	Work	Agile methods
Business analyst	17	30%	70%	0%	4	2
Scrum master	16	25%	69%	6%	15	3
Developer	14	71%	29%	0%	4	2
Product owner/product manager	7	29%	71%	0%	16	3
Tester/QA	6	33%	67%	0%	7	3
Agile coach	6	50%	50%	0%	20	7
Project manager	2	50%	50%	0%	8	4
Data scientist	2	50%	50%	0%	2	2
DevOps lead/architect	1	100%	0%	0%	6	5
Junior technician	1	100%	0%	0%	2	1
N/A	5	N/A	N/A	N/A	N/A	N/A
All	77	40%	53%	6%	9	3

Among the participants, there were five respondents who didn't report their personal information (e.g. role, gender, experience). However, we included their answers as they were complete otherwise.

Data Collection. All respondents received an eight-page online survey with an anonymous link distributed through social media posts and emails. The average time to complete the survey was about 8–10 min.

The entry page of the survey contained an introduction to the focus of the research, the target audience, duration, a confidentiality statement, and contact information. The first section included the research measures with several items. The second section of the survey collected demographic information. This included gender, organization name, total work experience and experience with agile practices in years, percentage of female participation in their teams and job title.

Measures. Existing validated survey items from literature were used to measure the selected variables. All items for the selected variables were measured on a five-point Likert scale ranging from 1 (Strongly Disagree) to 5 (Strongly Agree). While many research instruments are available to assess teamwork quality in agile teams (e.g. [4]), we applied the model of Hoegl and Gemuenden [7] due to its wide application, validity and inclusion of team performance measures.

Gender Diversity as a Headcount: We measured the gender diversity as a headcount using the Blau Index [27] which was defined as $(1 - (m^2 + f^2))$, where m denotes the fraction of males and f denotes the fraction of female team members. The Blau Index is a well-established diversity measure for categorical variables. However, in this study, we consider the perfect diversity as 50%. Thus, we reorganized this formula by dividing it by 0,5 to obtain the largest index 1 for 50% gender participation.

Organizational Support for Women's Advancement: Jawahar and Hemmasi [1] developed an instrument to measure the perceptions of employees for organizational support for women's advancement with 12 items. Three of the items were found to be cross-loaded in the three-factor solution of Jawahar and Hemmasi [1] and those items were eliminated in the hypothesis testing. Hence, the nine-item version of the scale was adapted for this study.

Communication: The quality of communication was measured by five items, based on Hoegl and Gemuenden [7]. Specifically, we adapted the Liang et al. [45] version for this study, which is a shorter and consolidated version of the scale for the ease of the respondents. Questions included statements about the frequency of conversations, spontaneity of conversations, and the perceived efficiency of communication between team members.

Coordination: Team-level coordination was measured by three items, adapted from Hoegl and Gemuenden [7]. Participants were asked about their perceptions for their team's capabilities to develop and agree upon a common task-related goal structure that has sufficiently clear sub-goals for each team member.

Balance of Member Contributions: The three-item measure was adapted from the work of Hoegl and Gemuenden [7] to understand the contribution of the task-relevant knowledge and experience of all members to the decision-making processes of the team.

Mutual Support: The quality of mutual support was measured by five items also adapted from Hoegl and Gemuenden [7]. Two of the items from the original scale were removed in order to increase the internal consistency and avoid the repetition of questions with similar meanings.

Effort: It is important that interaction between members minimizes social loafing and instead promotes a shared commitment among members to the team and its work. To measure the quality of the effort at a team level, the respondents answered three items adapted from Hoegl and Gemuenden [7]. One of the items from the original scale, that is "the team put(s) much effort into the teamwork" was removed due to its similarity with the first item, "Every team member fully pushes the teamwork."

Cohesion: The team-level quality of cohesion is measured with items adapted from the work of Hoegl and Gemuenden [7]. The original scale included ten items. However, to reduce the number of items to achieve a more aggregated scale, two of three similar items were dropped.

Team Agility: To measure the flexible capability of responding to unpredicted environmental changes in a timely manner, Liu et al.'s [43] measure with eight items were adapted for this study. Participants were asked about their team's reflexes in adapting to new skills and answering to changes in customer needs and organizational conditions. However, after discussion with participants, we observed that two of the original items were perceived differently in their meanings. Therefore, we eliminated them and proceeded with the six-items.

Team Performance: Hoegl and Gemuenden [7] described team performance in terms of effectiveness and efficiency. Effectiveness refers to the degree to which the team meets the expectations regarding quality of the outcome. In case of software development projects, effective performance is regularly attributed to predefined qualitative properties of the product to be developed, e.g. functionality, robustness, reliability, performance, etc. On the other hand, the team's efficiency is assessed in terms of adherence to schedules and budgets. In order to assess the team-level performance with the self-assessment, we adapted Hoegl and Gemuenden's [7] scale which included items.

Data Analysis. Before starting the analyses, we did preparatory cleaning of the data and several pre-processing analyses to show the descriptive statistics of the investigated variables and the correlations between them (Table 2). After finishing this organization of the data, we imported the Excel file into Stata

Table 2. Correlations between investigated variables (*p < 0.05)

	Variable	1	2	3	4	5	6	7	8	9	10	Alpha
1	Gender diversity as a headcount	1,00										N/A
2	Organizational support for women's advancement	0,26*	1,00									0,88
3	Communication	0,14	0,33*	1,00								0,90
4	Coordination	0,29*	0,37*	0,74*	1,00							0,85
5	Balance of member contributions	0,41*	0,43*	0,52*	0,67*	1,00						0,58
6	Mutual support	0,25	0,64*	0,63*	0,63*	0,74*	1,00					0,85
7	Effort	0,20	0,48*	0,40*	0,53*	0,62*	0,65*	1,00				0,79
8	Cohesion	0,27	0,64*	0,57*	0,55*	0,77*	0,80*	0,71*	1,00			0,91
9	Team agility	0,33	0,63*	0,61*	0,58*	0,69*	0,76*	0,60*	0,78*	1,00		0,91
10	Team performance	0,23	0,63*	0,54*	0,60*	0,74*	0,79*	0,65*	0,79*	0,84*	1,00	0,94

software where we conducted all the analyses for this research. Tables 3 and 4 show the detailed statistics for the ten variables and the additional three demographic variables included as control variables, years of working experience, years of working with agile methodologies and gender. Each variable is represented as one-factor solution of the individual items that comprise the variable. The factor analysis of the items is further discussed in the next section. Because of the standardized factors, the calculated means of the variables were so small, so we approximated them to zero.

Cronbach's alpha coefficients are also reported in Table 4. Each coefficient was calculated by including the measurement items of the respective variable. Nunnally and Bernstein [46] consider a Cronbach's alpha greater than 0.7 as satisfactory. All variables were thus satisfactory, except balance of member contribution which had an alpha value of 0.58. This is the same Cronbach's alpha value for the balance of member contributions reported in the Lindsjrn et al.'s [8] study.

Reliance on self-reported data can create concern about common source bias [47]. Doty and Glick [48] reported that the bias is typically not large enough to affect theoretical interpretations of substantive relationships. Nonetheless, Harman's one-factor test [49] was used to assess the extent to which intercorrelations among the variables might be an artefact of common source bias. The test requires all items to be entered into a factor analysis. The basic assumption of the technique is that if a substantial amount of common source bias is present, either a single factor emerges from the factor analysis or one general factor accounts for the majority of the variance [47]. Accordingly, all 54 items were factor analyzed. The analysis resulted in a nine-factor solution as evidenced by the Scree test. Hence, no general factor emerged from the analysis. The result of the test indicated that common source bias is not a serious threat. This interpretation is consistent with the conclusion of [50] in which the authors reported the bias as existing only at low and usually inconsequential levels.

5   Results

In this study, cross-sectional regression analyses were conducted. However, first we predicted the scores of the variables using factor analysis for each participant. Furthermore, in Phase I and Phase II of the conceptual model, the hypotheses were tested with multivariate regression analysis for each proposed hypothesis. In this chapter, the prediction of variables and the results of the regression tests are discussed.

5.1   Factor Analysis

Prior to testing the conceptual model, factor analysis was conducted to predict the variables measured with multi-item scales. The purpose of the factor analysis is to identify the lowest possible number of constructs needed to reproduce the original data [51]. The desired result for this study, therefore, was that all items measuring the same variable are scored together to give the best estimate of each participant's score on that variable. All items of each of the nine variables was factor-analyzed and following the Kaiser criterion, the eigenvalues above 1.0 and the scree plot suggested a one-factor unrotated solution for each variable. Table 3 provides the eigenvalues of the one-factor solutions and the percentage of variance explained.

Table 3. One - factor solution of the variables, eigenvalues and percentage of variables explained

	One-factor solution	
Variable	Eigenvalue	% of variance explained
Organizational support for women's advancement	4,66	51,8%
Communication	3,19	63,8%
Coordination	1,86	61,9%
Balance of member contributions	1,06	35,5%
Mutual support	2,64	52,9%
Effort	1,67	55,6%
Cohesion	4,18	59,8%
Team agility	3,85	64,1%
Team performance	6,63	60,3%

5.2   Regression Results

To test the hypotheses in Phase I, regressions were conducted including the gender as control variable. According to results reported in Table 4, while gender diversity as a headcount has no significant impact on communication, mutual

Table 4. Summary of regression analysis for each dependent variable

Dependent variables	Gender diversity as a headcount			Organizational support for women's advancement			R-squared
	β	SE	p-value	β	SE	p-value	
Communication	−0,23	0,64	0,57	0,29	0,13	0,03	0,13
Coordination	0,10	0,56	0,04	0,29	0,11	0,01	0,19
Balance of member contributions	1,18	0,44	0,00	0,37	0,09	0,00	0,36
Mutual support	−0,40	0,47	0,36	0,61	0,09	0,00	0,46
Effort	−0,23	0,53	0,07	0,48	0,11	0,00	0,28
Cohesion	0,49	0,52	0,21	0,60	0,10	0,00	0,42
Team agility	0,84	0,50	0,10	0,60	0,10	0,00	0,45

support, effort, cohesion and team agility, it has positive and significant impact on coordination and balance of member contributions at 5% level and 1% level respectively. Hence, only H2 and H3 were supported among the hypotheses. Interestingly, beta values of gender diversity as a headcount for communication, mutual support, and effort were negative. However, based on their p-values, we can say that there is no significant evidence to support a negative impact of gender diversity as a headcount on these variables. On the other hand, organizational support for women's advancement has a significant positive impact on the balance of member contributions, mutual support, effort, cohesion, and team agility at 1% level and a positive and significant impact on communication and coordination at 5% level. Thus, all hypotheses regarding the organizational support for women's advancement were supported. Furthermore, on each regression, the control variable gender had insignificant effects.

In the second phase of the model, we tested if communication, coordination, balance of member contributions, mutual support, effort, cohesion, and team agility had a significant impact on team performance by adding the total years of experience and the years of experience with agile as control variables. As we discussed, the first six variables were defined as teamwork quality sub-constructs in Hoegl and Gemuenden's study [7]. However, in this study, we are concerned with their individual effects since they had high correlation coefficients among each other as demonstrated in Table 2. In each regression, we added the total years of experience and the years of experience with agile as control variables.

In each regression, the teamwork quality variables and the team agility had a significantly positive impact on team performance at 1% level, thus supported the all hypotheses in Phase II. Moreover, except in the regression with team agility, the control variable, years of experience with agile methodologies had a positive and significant impact on all regressions at 5% level, but the total years of experience had no significant impact in all regressions.

6 Discussion

Our results, as summarized in Fig. 1, indicate that the organization's support for women's advancement has a stronger impact on teamwork quality and eventually team performance, than gender diversity as a headcount only. While our results imply that team-level coordination and balance of member contributions is affected by gender diversity at ($p < 0.05$), organizational support for women's advancement has a positive correlation with all of the TWQ dimensions at $p < 0.01$.

This indicates that while organizational policies to support career growth of women have a positive impact on teamwork quality and indirectly on performance through TWQs, the impact of gender composition on team performance is much more contextual. This is in line with previous studies that concluded that an equal team composition of males and females does not automatically lead to better performance. Rogelberg and Rumery [14] studied five gender compositions (all-male, lone-female, balanced-gender, lone-male, and all-female teams), and found that for a male-oriented task (winter survival exercise), lone-female teams outperformed all other gender compositions in decision quality, time on task and cohesion. Experimental studies in business settings indicate groups of two men and one women were most successful [15]. Turban et al. [13] found that the effect of gender diversity on performance varies significantly across countries and industries due to differences in institutional contexts.

In the following subsections we will further discuss our findings in line with the existing literature.

6.1 Organizational Support for Women's Advancements vs. Gender Diversity as a Head Count

While perceived organizational support (POS) for women's advancement has been previously established as a strong predictor of job satisfaction and turnover intentions [1], with previous studies pointing to a positive impact of POS [1], or the effect of human resource management practices promoting equal career opportunities and work-family integration [52] on the turnover intentions of female managers (intentions to leave an organization), our findings indicate its impact on team-level performance in agile teams.

One interpretation of this finding could be that the organizational support for women's advancement affects the cultural dimensions of an organization and creates a longer lasting effect rather than a practical measure or even coincidental fact of having a gender diverse team, as perceived by our participants. Previous research as well as agile practitioner frameworks, reiterate the importance of an organizational culture being compatible with agile methods for those to be effective in context [53]. In that sense our findings could be understood that the support for women's advancement being in line with the broader agile mindset, which might make it interesting to look at further effects within the organization beyond teamwork quality and performance.

These results are quite interesting and in a way show the nature of agile methodologies - the mindset of agile enables the members of software development teams to explicitly embrace teamwork qualities just as much as the product development [3]. Hence, for the teamwork qualities of the unsupported hypotheses, an equal gender involvement might not be necessary as it is an integrated part of the agile teamwork.

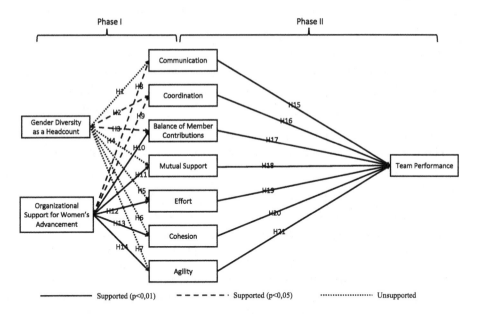

Fig. 1. Resulting research model

There can be several reasons for not observing a significant relationship between gender diversity and communication. As we have already mentioned, Razavian and Lago's [10] study suggested that a strong level of communication could be attributed to feminine expertise in software development teams which would mean assigning holders of feminine expertise to activities to which they could really contribute. In this study, we defined the perfect diversity as 50% female representation. Therefore, these results may suggest that to promote communication through feminine communication skills, reaching an equal gender head count might not be necessary to empower communication. A similar result was also reached by the study of Catolino et al. [6] in which they found that Organizational Silo, defined as a common community smell in software development teams as a result of ineffective communication, was not significantly reduced with equal gender participation ($p < 0.1$).

We observed that both gender diversity and organizational support for women's advancement have a significant relationship to coordination. Combining this observation with Weileman and Brune's [24] suggestion that female leadership provides an advantage for enhanced team coordination, it can be said that coordinating skills of females becomes more effective when they are involved in agile teams with higher equal representation. Moreover, being in an organizational culture where equal opportunities are provided, a high quality of coordination in agile teams can be expected.

The results of gender diversity for the team-level balance of member contributions is consistent with the argument made by Weilemann and Brune [24], which discussed female leadership as being more democratic and encouraging a more shared decision-making environment. In addition, in an organization where the advancement of female employees is supported, a better balance of member contributions can be expected.

We developed H4 based on the findings of Eagly and Johannessen-Schmidth [12] on the superior empathetic behavior of females. However, considering the results, it can be suggested that for women to promote mutual support in agile teams, equal participation of men and women is not necessary but the organizational support for women's advancement is important.

The contradictory findings of gender diversity in relation to effort with the proposed hypothesis can be linked to the conclusion that an equal gender composition is not necessary for team members to put their full effort into pushing projects in agile teams. However, it is clear that perceiving the inclusive attitude of organizations who support women's advancement may be necessary to generate sufficient effort in agile teams.

In accordance with the results, to improve cohesion in agile teams, increasing gender diversity by means of headcount is not important. Our expectation was that social orientation of females would be more influential and thus enhance the interpersonal cohesion among agile team members; however, we concluded with a similar result to the findings of Rogelberg and Rumery [14]. In their study of interpersonal cohesion conducted among psychology students, no significant differences were observed between gender-diverse and lone female/male teams.

Team agility is not affected by an increase in gender diversity but it can be higher in organizations where females' career development is supported within the organization. As a consequence, these findings partially confirm the evidence in the existing literature [24] that agile development may particularly benefit from an organizational culture that promotes equality.

6.2   Linking Teamwork Quality to Team Performance

In phase II of the study, we tested the impact of teamwork quality variables and the team agility variable on individual team performance (see Fig. 1), thus, comparing our data to the previously established results of Hoegl and Gemuenden [7] and Lindsjrn et al. [8] applying the Teamwork Quality model.

Each variable has a significantly positive impact on team performance in agile software development teams. In the studies by Hoegl and Gemuenden [7] and Lindsjørn et al. [8], these individual variables were used to construct teamwork quality as a higher-order construct and its significant impact on team performance at 1% level was demonstrated, leading us to expect significant relations between its sub-constructs and team performance. Moreover, in the study by Liu et al. [43], the significant impact of team agility on team performance was also reported at 1% level.

Thus, to achieve high-quality teamwork and high performance, organizations can intervene by increasing diverse gender representation and enhancing support for career development by adopting policies, procedures and programs to engender support within the agile teams and the organization as a whole. This might require a mindset shift not only at the team level but in all levels of the organization. Support for the advancement of women could be improved by overcoming barriers (e.g. good old-boy networks, subconscious bias), treating men and women equally, providing mentoring relationships, increasing women's access to important networks and monitoring promotion decisions. Such adjustments might also influence male employees' attitudes towards the organization; our survey found that male respondents were aware of support mechanisms within the organization and able to pass accurate judgment on it. From the results, it seems that organizations' support for women's advancement can be a strong determinant of both teamwork quality and team agility variables. Moreover, as we disucssed in relation to Table 2, organizations' support for women's advancement has a greater correlation and significant correlation with team performance (0,63) while gender diversity has an insignificant and thus weak correlation (0,23) with it.

6.3 Potential Limitations and Directions for Future Research

Potential limitations of this research should be recognized to pave the way for any future studies. First of all, due to the small-sized data set containing the responses of 77 professionals, conclusions have had to be carefully drawn. We have therefore, paid particular attention to the quality of our collected data by encouraging participants to give realistic answers and emphasizing the anonymous treatment of data to establish a reasonable level of trust. However, given the limited number of answers, we were unable to cluster the sample based on country, title, or sector which could have provided detailed results for comparison. Thus, for future research, we would advise that the data set be expanded to increase the generalizability of the findings and to decrease the risk of selective effect bias.

Another limitation could be that this study is cross-sectional, and this may limit the ability to achieve causal inferences from the data. Longitudinal studies are needed if gender diversity or organizational support for women's advancement changes over time, and to establish whether they are of significant influence on teamwork quality variables and agility. In addition, these could provide insights into differences between teams that are working with agile methodologies over a longer period of time and those that have just been established. It would be interesting to examine whether the teamwork qualities, agility, and performance change over time with different projects.

6.4 Conclusion

In this paper we have presented the results of our quantitative study to explore the impact of gender equality on agile software development teams. In particular, by incorporating the two aspects of gender equality in the teamwork quality and performance model of Hoegl and Gemuenden [7] we have analyzed the impact of gender diversity and organizations support for the advancement of women on teamwork quality and degree of agility, and further tested the role of teamwork quality and agility on team performance.

Our study suggests that gender equality and its impact on team performance is not strictly related to a 50/50 gender composition of teams, but that organizational policies that support the advancement of women, and the perceptions of these policies do have an impact on teamwork quality (TWQ) and team performance. Specifically, the findings imply that policies and organizational structures providing equal career opportunities and equal respect have an impact on teamwork quality and eventually team performance. By supporting the advancement of female employees, organizations may improve the quality of communication, coordination, balance of member contributions, mutual support, effort, cohesion, and agility.

Such efforts may need to be coordinated with efforts to overcome the existing barriers standing in the way of women's equal participation in agile teams. Of course, this investigation is still strongly limited due to the sample size, but we consider these concerns worth studying further. To the best of our knowledge this is the first work that provides empirical evidence that teamwork quality and agility of agile software development teams are affected by both gender diversity and organizations' support for women's advancement. With this paper we hope to encourage other researchers to build up research in this area and further develop strategies to promote the inclusion of females.

Acknowledgments. The authors would like to thank the survey participants and anonymous reviewers for generously contributing to this study.

References

1. Jawahar, I., Hemmasi, P.: Perceived organizational support for women's advancement and turnover intentions. Women Manag. Rev. **21**, 643–661 (2006)
2. Hoda, R., Noble, J., Marshall, S.: Organizing self-organizing teams. In: Proceedings of the 32nd ACM/IEEE International Conference on Software Engineering, vol, 1, pp. 285–294 (2010)
3. Whitworth, E., Biddle, R.: The social nature of agile teams. In: Agile 2007 (AGILE 2007), pp. 26–36. IEEE (2007)
4. Stettina, C.J., Heijstek, W.: Five agile factors: helping self-management to self-reflect. In: OConnor, R.V., Pries-Heje, J., Messnarz, R. (eds.) EuroSPI 2011. CCIS, vol. 172, pp. 84–96. Springer, Heidelberg (2011). https://doi.org/10.1007/978-3-642-22206-1_8
5. Vasilescu, B., et al.: Gender and tenure diversity in Github teams. In: Proceedings of the 33rd annual ACM Conference on Human Factors in Computing Systems, pp. 3789–3798 (2015)
6. Catolino, G., Palomba, F., Tamburri, D.A., Serebrenik, A., Ferrucci, F.: Gender diversity and women in software teams: how do they affect community smells? In: 2019 IEEE/ACM 41st International Conference on Software Engineering: Software Engineering in Society (ICSE-SEIS), pp. 11–20 (2019)
7. Hoegl, M., Gemuenden, H.G.: Teamwork quality and the success of innovative projects: a theoretical concept and empirical evidence. Organ. Sci. **12**(4), 435–449 (2001)

8. Lindsjørn, Y., Sjøberg, D.I., Dingsøyr, T., Bergersen, G.R., Dybå, T.: Teamwork quality and project success in software development: a survey of agile development teams. J. Syst. Softw. **122**, 274–286 (2016)
9. Bear, J.B., Woolley, A.W.: The role of gender in team collaboration and performance. Interdisc. Sci. Rev. **36**(2), 146–153 (2011)
10. Razavian, M., Lago, P.: Feminine expertise in architecting teams. IEEE Softw. **33**(4), 64–71 (2015)
11. Werner, L.L., Hanks, B., McDowell, C.: Pair-programming helps female computer science students. J. Educ. Res. Comput. (JERIC) **4**(1), 4-es (2004)
12. Eagly, A.H., Johannesen-Schmidt, M.C.: The leadership styles of women and men. J. Soc. Issues **57**(4), 781–797 (2001)
13. Turban, S., Wu, D., Zhang, L.: When gender diversity makes firms more productive (2019)
14. Rogelberg, S.G., Rumery, S.M.: Gender diversity, team decision quality, time on task, and interpersonal cohesion. Small Group Res. **27**(1), 79–90 (1996)
15. Apesteguia, J., Azmat, G., Iriberri, N.: The impact of gender composition on team performance and decision making: evidence from the field. Manag. Sci. **58**(1), 78–93 (2012)
16. Accenture: Creating a culture of equality in the workplace (2020)
17. McKinsey: Women in the workplace 2020 (2020)
18. Hopkins, M.M., O'Neil, D.A., Passarelli, A., Bilimoria, D.: Women's leadership development strategic practices for women and organizations. Consult. Psychol. J. Pract. Res. **60**(4), 348 (2008)
19. Eagly, A.H., Johnson, B.T.: Gender and leadership style: a meta-analysis. Psychol. Bull. **108**(2), 233 (1990)
20. Beranek, G., Zuser, W., Grechenig, T.: Functional group roles in software engineering teams. In: Proceedings of the 2005 Workshop on Human and Social Factors of Software Engineering, pp. 1–7 (2005)
21. Blum, L., Frieze, C., Hazzan, O., Dias, M.B.: A cultural perspective on gender diversity in computing. Reconfiguring the firewall, pp. 109–129 (2007)
22. Dubinsky, Y., Hazzan, O.: A framework for teaching software development methods. Comput. Sci. Educ. **15**(4), 275–296 (2005)
23. Katzy, B.R., Stettina, C.J., Groenewegen, L.P., de Groot, M.J.: Managing weak ties in collaborative work. In: 2011 17th International Conference on Concurrent Enterprising, pp. 1–9. IEEE (2011)
24. Weilemann, E., Brune, P.: Less distress with a scrum mistress? On the impact of females in agile software development teams. In: Proceedings of the ASWEC 2015 24th Australasian Software Engineering Conference, pp. 3–7 (2015)
25. Ruderman, M.N., Ohlott, P.J.: Leading roles: what coaches of women need to know. Leadersh. Action Publ. Center Creative Leadersh. Jossey-Bass **25**(3), 3–9 (2005)
26. Hewlett, S.A., et al.: The athena factor: reversing the brain drain in science, engineering, and technology. Harvard Bus. Rev. Res. Rep. **10094**, 1–100 (2008)
27. Blau, P.M.: Inequality and Heterogeneity: A Primitive Theory of Social Structure, vol. 7. Free Press, New York (1977)
28. Myaskovsky, L., Unikel, E., Dew, M.A.: Effects of gender diversity on performance and interpersonal behavior in small work groups. Sex Roles **52**(9–10), 645–657 (2005)
29. Ely, R.J., Thomas, D.A.: Cultural diversity at work: the effects of diversity perspectives on work group processes and outcomes. Adm. Sci. Q. **46**(2), 229–273 (2001)

30. Carli, L.L.: Gender and social influence. J. Soc. Issues **57**(4), 725–741 (2001)
31. Craig, J.M., Sherif, C.W.: The effectiveness of men and women in problem-solving groups as a function of group gender composition. Sex Roles **14**(7–8), 453–466 (1986)
32. Taps, J., Martin, P.Y.: Gender composition, attributional accounts, and women's influence and likability in task groups. Small Group Res. **21**(4), 471–491 (1990)
33. Mirvis, C.: Lawer (1977) job satisfaction and job performance in bank tellers. J. Soc. Psychol. **133**(4), 564–587 (1980)
34. Shaw, M.E.: Group Dynamics: The Psychology of Small Group Behavior. McGraw-Hill College, New York (1981)
35. Cartwright, D.: The nature of group cohesiveness. Group Dyn. Res. Theory **91**, 109 (1968)
36. Pikkarainen, M., Haikara, J., Salo, O., Abrahamsson, P., Still, J.: The impact of agile practices on communication in software development. Empirical Softw. Eng. **13**(3), 303–337 (2008)
37. Mackenzie, A., Monk, S.: From cards to code: how extreme programming re-embodies programming as a collective practice. Comput. Support. Coop. Work (CSCW) **13**(1), 91–117 (2004)
38. Hackman, J.R.: Groups that work and those that don't. Number E10 H123. Jossey-Bass (1990)
39. Tjosvold, D.: Cooperative and competitive goal approach to conflict: accomplishments and challenges. Appl. Psychol. **47**(3), 285–313 (1998)
40. Mudrack, P.E.: Defining group cohesiveness: a legacy of confusion? Small Group Behav. **20**(1), 37–49 (1989)
41. Anderson, L.R., Blanchard, P.N.: Sex differences in task and social-emotional behavior. Basic Appl. Soc. Psychol. **3**(2), 109–139 (1982)
42. Christopher, M.: The agile supply chain: competing in volatile markets. Ind. Mark. Manag. **29**(1), 37–44 (2000)
43. Liu, M.L., Liu, N.T., Ding, C.G., Lin, C.P.: Exploring team performance in high-tech industries: future trends of building up teamwork. Technol. Forecast. Soc. Change **91**, 295–310 (2015)
44. Conrad, L.: Employee empowerment in services: a framework for analysis. Pers. Rev. **28**(3), 169–191 (1999)
45. Liang, T.P., Wu, J.C.H., Jiang, J.J., Klein, G.: The impact of value diversity on information system development projects. Int. J. Project Manag. **30**(6), 731–739 (2012)
46. Nunnally, J.C.: Psychometric Theory 3E. Tata McGraw-Hill Education, New York (1994)
47. Podsakoff, P.M., MacKenzie, S.B., Lee, J.Y., Podsakoff, N.P.: Common method biases in behavioral research: a critical review of the literature and recommended remedies. J. Appl. Psychol. **88**(5), 879 (2003)
48. Doty, D.H., Glick, W.H.: Common methods bias: does common methods variance really bias results? Organ. Res. Methods **1**(4), 374–406 (1998)
49. Podsakoff, P.M., Organ, D.W.: Self-reports in organizational research: problems and prospects. J. Manag. **12**(4), 531–544 (1986)
50. Keeping, L.M., Levy, P.E.: Performance appraisal reactions: measurement, modeling, and method bias. J. Appl. Psychol. **85**(5), 708 (2000)
51. Gorsuch, R.L.: Exploratory factor analysis: its role in item analysis. J. Pers. Assess. **68**(3), 532–560 (1997)

52. Nie, D., Lämsä, A.M., Pučėtaitė, R.: Effects of responsible human resource management practices on female employees' turnover intentions. Bus. Ethics Eur. Rev. **27**(1), 29–41 (2018)
53. Iivari, J., Iivari, N.: The relationship between organizational culture and the deployment of agile methods. Inf. Softw. Technol. **53**(5), 509–520 (2011)

The State of Agile Software Development Teams During the Covid-19 Pandemic

Krzysztof Marek[1]([✉]) [iD], Ewelina Wińska[2] [iD], and Włodzimierz Dąbrowski[1] [iD]

[1] Warsaw University of Technology, 00-661 Warsaw, Poland
krzysztof.marek@pw.edu.pl
[2] Polish-Japanese Academy of Information Technology, 02-008 Warsaw, Poland

Abstract. The Covid-19 pandemic in 2020 forced Agile Software Development Teams (ASDT) to rapidly transition to remote work and adapt to new business circumstances. The focus of this research was to investigate the impact of the Covid-19 pandemic on ASDT work and what tools and metrics are used by ASDT. A global survey was performed with 120 answers from different software engineering teams. The results of the research indicate that the work of ASDT wasn't significantly impacted. Most of the ASDT had experience with working in a distributed or remote environment. Therefore, most of the ASDT were able to transitioned to full remote work. Results indicate the Covid-19 pandemic didn't have much impact on Product Backlog and Vision. Moreover, most ASDT didn't change their Definition of Done and release frequency, indicating that the pace and quality of work wasn't disturbed during the Covid-19 pandemic. The few ASDT that changed their work organization did it together with changes to Product Backlog and Vision. Results indicate that the prevalence of distributed teams and remote work among ASDT helped with the transition to fully remote work during the Covid-19 pandemic. Additionally, this article presents gathered data of popularity of different online cooperation tools and metrics used by ASDT.

Keywords: Agile software development · SAFe · LeSS · Scrum · Kanban · Collaboration tools · Agile metrics · Distributed teams · Remote work · Survey · Covid-19

1 Introduction

Currently the different Agile approaches are used worldwide to develop software, with distributed Agile teams becoming more and more common. A recent study performed at the end of 2019 by VersionOne on "The State of Agile" [1] reports that 95% of interviewed companies use agile development methods with 51% respondents stating that it is used in more than half of their teams.

The study performed by Sharma et al. [2] indicates that the Scrum framework is the most popular of all Agile frameworks and methodologies both in industrial use and in scientific research. According to Sharma's research Scrum is constantly gaining popularity in the industry. Many teams have adopted the Scrum framework. This naturally led to the scaling up of Scrum or other frameworks, as well as their adaptation to

© Springer Nature Switzerland AG 2021
A. Przybyłek et al. (Eds.): LASD 2021, LNBIP 408, pp. 24–39, 2021.
https://doi.org/10.1007/978-3-030-67084-9_2

distributed teams. However, introducing Agile practices to a distributed team requires overcoming multiple communication obstacles [3] and creating a transformation strategy [4]. The initial agile frameworks like Scrum or Extreme Programing (XP) were created for small, co-located teams. Teams small size and co-location facilitates communication, cooperation, self-organization and allows for quick reactions to rapid changes on the market. However, as globalization progressed, distributed teams started becoming a worldwide standard. In order to still benefit from the advantages provided by Agile frameworks and methodologies the practices needed to be adjusted to the new characteristic of distributed teams. Such transformations were already successful in the past [5, 6], usually Agile Software Development Teams (ASDT) were using a mixed approach in order to facilitate communication, increase transparency and reinforce feedback loops in distributed environment.

The core of these mixed strategies were online tools. Their introduction allowed for maintaining communication and knowledge sharing between distributed team members [7]. However, communication facilitation is not sufficient on its own in distributed teams. The transparency of teams' work is significantly reduced in distributed environment. The initial solution was to introduce tools to visualize tasks and to track everyone's work [8, 9]. Such solutions worked, however they turned out to be insufficient for more matured ASDT. These teams and organizations started to introduce different metrics [10], customized to the individual characteristic of the team and organization. Today, due to automatization and the use of online tools, such metrics sourced additional information from already existing data, without impacting team members' every-day work.

Therefore, the best results can be achieved by the use of both communication tools and metrics, as they complement each other. By using both, the team can easily communicate, visualize current work and observe their progress, effectiveness, quality of the product and distribution of effort. This enables the ASDT to make data driven decisions at any time.

1.1 Problem Statement

In early 2020 the global Covid-19 pandemic started. Multiple countrywide lockdowns and market uncertainty forced small [11] and large [12] organizations to reevaluate their business plans. Moreover, all the software development teams were forced to start working from home, creating an additional challenge for management and teams to organize remote work in a very short time. This was an unprecedented situation. All the teams, almost instantaneously had to start working remotely, making every team distributed at least within a single country.

1.2 Objective

The work presented in this paper aims to build an initial understanding of Covid-19's influence on ASDT' organization of work. The objective of this work is to determine what metrics and tools are used by ASDT and how the Covid-19 pandemic impacted the work organization of ASDT. The following research questions were created:

- How the ASDT responded to the circumstances of the Covid-19 pandemic?
- What tools and metrics are used by ASDT?

1.3 Contribution

For the purpose of this study a total of 120 answers from different Agile software development practitioners were examined. The respondents fulfilled different roles from regular team members to C-level management and came from a wide spread of industries and organization sizes. The survey consisted of questions investigating the characteristic of the organization, the impact of Covid-19 on the teams' work and what tools and metrics are used.

1.4 Overview

In the second chapter an overview of related works was presented. The third chapter describes the research design and methodology. The forth chapter presents the survey results and was divided into three subsections. The first subsection presents the respondents characteristics, the second section describes the Covid-19 pandemic impact on ASDT, the last subsection presents the tools and metrics used by ASDT. The fifth chapter presents the discussion of the survey results and indicates possible future work. The last chapter contains the conclusions of the survey study.

2 Related Work

Not much research has been published describing the Covid-19 pandemic's influence on ASDT as the issue is new. In the history of software development there is no precedent for such a forced, rapid, global, industrywide move to remote work. A recent survey performed by Raišienė et al. [13] pictures the influence of rapid introduction of remote work, also known as telework, on Lithuanian workers in many different occupations. However, these interesting findings don't shed much light on the situation of ASDT and how the tools and practices from distributed teams helped with the rapid transition to remote work.

The possible impact of Covid-19 on Agile was discussed by Mancl et al. [14] in his article based on a panel discussion during the XP2020 conference. Based on their experience they emphasize the importance of carefully selected online tools facilitating the communication and self-organization of the ASDT. The possibility of simulating conditions similar to an in-person meeting with a whiteboard is described as critical for the success of an Agile team. The importance of online telecommunication tools: text, audio and visual in distributed ASDT was brought up in an article by Robinson [6]. As described by Mancl et al. [14] proper use of tools turned out to be crucial when all teams became distributed.

The use of metrics in software development has been a subject of research for a long time. A few years before the Agile Manifesto was signed Schwaber [15] puts emphasis on the importance of measurements in empirical process, the base of Scrum framework. In this work the need for the development of metrics for empirical processes was indicated. Later Hartmann et al. [16] stressed the benefits of measuring Velocity in Scrum projects and proposed a set of additional useful metrics. Metrics can deliver additional information for decision making and monitoring without putting a constrain of ASDT work, therefore Downey et al. [10] proposed a set of metrics for fast working ASDT. Ladas in his book "Scrumban" [17] proposed to use elements of Kanban in ASDT using Scrum as a way to support the software development process and enable ASDT to transition to Kanban in the future. Anderson in his book [18] describes a set of Agile metrics inspired by Toyota Production System [19] as a core of the Kanban Method. Literature studies performed in recent years by Kupiainen et al. [20] and Kurnia et al. [21] indicate that ASDT use different metrics in their work and measuring different aspects of Agile software development is becoming a standard practice.

The state of Agile practices before the Covid-19 pandemic in different teams was well described in the "State of Agile" industry survey performed by VersionOne [1]. This survey was performed between August and December 2019 and gathered 1121 responses from around the world. The resulting report allows for a better understanding of the Agile practices in use, including the use of frameworks, tools and metrics. However only 63% of respondents work in Software Development or IT. Therefore, it provides an insight to all types of Agile teams, not specifically the ASDT.

3 Research Design and Methodology

For the purpose of the empiric study a survey was designed. The initial pool of questions was created by the authors, then the first version was reviewed by 4 independent Agile practitioners working as experts in international software development companies. The remarks to the first version were included in the final version. The final survey, composed of 22 questions with 18 closed-ended and 4 open-ended questions, was divided into four parts. The first nine questions characterized the participant by asking about their country of origin, role in their organization, level of teams' distribution, used frameworks, remote work pre and post the Covid-19 pandemic, as well as their organization's size, industry and type. The second group of questions investigated the impact of the Covid-19 pandemic on Product Backlog and Vision, changes in: stakeholders' involvement; release frequency and Definition of Done. The third group of questions collects information about used metrics and reasons behind their use. It also asks if any new metrics were introduced during the Covid-19 pandemic. The last questions ask about tools used by the teams.

The anonymous survey was created in Google Forms and distributed through a direct approach and social media channels including Facebook and LinkedIn researchers professional networks, Agile software development practitioners groups and pages. The responses were gathered from 01.09.2020 to 11.09.2020. A total of 120 answers were submitted during this period. No partial answer was submitted, because all close-ended questions were obligatory. During the answers inspection no obviously biased or fake answer was detected, therefore no answer was deleted or omitted. The results were exported from Google Forms and imported to Excel. With the use of a spreadsheet tool the data was explored and visual figures were generated.

4   Results

In this section the 120 results of the survey are presented. The first subsection presents an overview of respondents teams. The next subsection presents the influence of Covid-19 on ASDT work. The last subsection presents tools and metrics used by ASDT.

4.1   Teams Characteristic

The first group of questions was designed to characterize the surveyed organization and team. The first question asked about the frameworks and methodologies used in the project. Respondents could select multiple options, with many choosing to do so. As shown in Fig. 1 the most common framework was Scrum (108 answers, 90% of respondents), followed by Kanban (50; 41.7%), DSDM or AgilePM (10; 8.3%), SAFe (8; 6.7%), Nexus and LeSS (4; 3.3% each), XP (3; 2.5%) Scrum@Scale, LeanSD and Waterfall (2; 1.7% each). There were 4 other responses (3.3% of respondents) mentioning self-developed frameworks. The most commonly combined frameworks were Scrum and Kanban with 38 concurrent occurrences (31.7% of respondents). Kanban, despite being the second most popular framework, is mostly used together with other frameworks. Only 8 respondents used Kanban exclusively (6.7% of respondents, 16% of Kanban practitioners). On the other hand Scrum, the most popular framework, is used on its own by 48 respondents (40% of respondents, 44.4% of Scrum practitioners). Moreover, we can also divide Scrum into two categories: Scaled and Nonscaled Scrum. If we count scaled Scrum frameworks (Nexus, LeSS, SAFe, Scrum@Scale) as one it shows that 18 respondents scale Scrum (15% of respondents, 16.7% of Scrum practitioners). On the other hand Scrum is not scaled by 90 respondents (75% of respondents, 83.3% of Scrum practitioners).

The second question asked about the participant's country of origin. Respondents were from 14 different countries: Bulgaria, Canada, China, Denmark, France, Gibraltar, Hong Kong, India, Ireland, Poland, Singapore, Spain, the United Kingdom and the United States. Most of the respondents (85% of all results) were from Poland, the country where the research team was based.

The third question investigated the distribution of the team. As shown in Fig. 2, the most common continent was Europe with 94 answers (78.3% of all answers), then Asia with 39 answers (32.5%) and North America with 38 answers (31.7%). A total of 8 respondents had team members in Australia (6.7%), 2 in South America (1.7%) and 1 in Africa (0.8%). Off all the polled teams 21 (17.5%) were not distributed, 50 (41.7%) were distributed within a single continent, 23 (19.2%) were distributed across two continents, 19 (15.8%) were distributed across three continents and 7 (6%) were distributed across four or more continents as shown in Fig. 3.

The fourth and fifth question asked about remote work before and after the start of the Covid-19 pandemic. The results have been presented in Fig. 4. Before the pandemic exactly half of respondents were working in a mixed model, a few days remotely, a few days onsite. Only 10 (8.3%) of the respondents were working fully remotely and 50 (41.7%) of the respondents were working fully onsite. After the start of the pandemic no one was working fully onsite. The majority, in total 103 (85.8%) of the respondents, was working fully remotely. Only 17 (14.2%) of the respondents were working in a

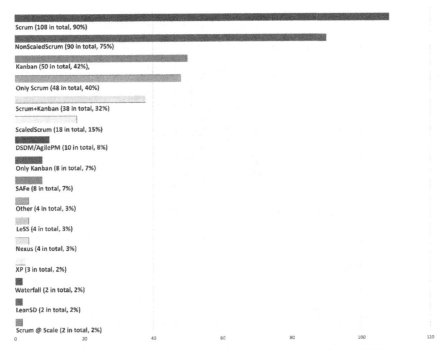

Fig. 1. Usage of different methodologies and frameworks in ASDT

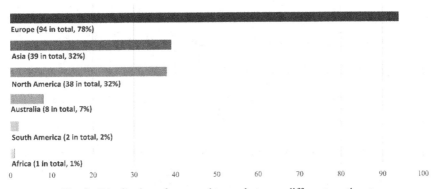

Fig. 2. Distribution of surveyed teams between different continents

mixed model, 8 of these 17 people used to work in a mixed model and 9 used to work fully onsite before Covid-19 pandemic. Therefore, from the 60 people that used to work in the mixed model 86.6% were able to transition into fully remote work. From the 50 people working only onsite, 82% were able to transition into full remote work, with the remaining 18% transitioning to a mixed model. All fully remote workers stayed fully remote.

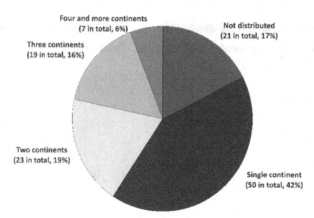

Fig. 3. Levels of surveyed teams' distribution

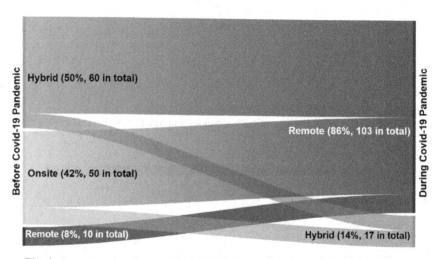

Fig. 4. Remote and onsite work in ASDT before and during the Covid-19 pandemic

In Fig. 5 the industries of the respondents have been presented. Respondents could select multiple answers. A wide spread of different industries can be observed. The most popular industries were "Financial Services, Banking & Insurance" and "High-tech, Electronics & Industrial Engineering" with 33 representatives each. The respondent's organizations size is also diverse. A total of 39 (32.5%) respondents work in an organization with more than 5000 employees. The other four categories were: 1–50, 51–300, 301–1000, 1001–5000. They each contained between 15.8% and 18.3% of respondents. Moreover, 16 respondents (13.3%) identify their organization as a start-up, with one employing over 5000, one 1001–5000, two 301–1000, two 51–300 and ten 1–50.

In the ninth question participants were asked to select roles they fulfil in the team. They could select multiple options. As shown in Fig. 6, the most common role was a Team Member with 47 answers (39.2%). The next two were Team Leader (23 answers, 19.2%)

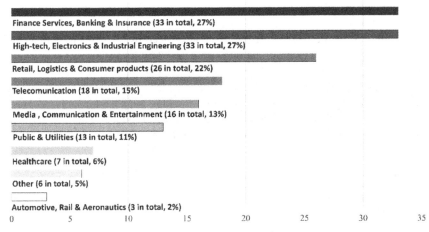

Fig. 5. Industries in which surveyed ASDT work

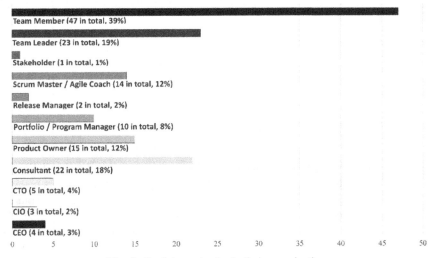

Fig. 6. Participants' roles in their organizations

and Consultant (22, 18.3%), then Product Owner with 15 representatives, Scrum Master/Agile Coach with 14 representatives, C-level with 12, Portfolio/Program Manager with 10, Release Manager with 2 and a single Stakeholder.

4.2 Pandemic Impact on ASDT Work

To measure the impact of the Covid-19 pandemic on the ASDT, the respondents were asked if the Covid-19 pandemic impacted the content of the Product Backlog or the Product Vision. Every respondent stated that they have easy access to the Product Backlog while working remotely, and therefore should be aware of any Covid-19 influence.

As shown in Fig. 7, 59 of the respondents (49.2%) stated that both the Product Backlog and the Product Vision were not impacted. In 16 cases the Product Backlog was not impacted, despite the Product Vision being influenced by the pandemic. In a single case it was a significant impact, in the other 15 cases Product Vision was only slightly impacted. In 37 cases the Product Backlog was slightly impacted, in 12 of these cases the Product Vision was not impacted and in the other 25 cases the Product Vision was slightly impacted. A drastic change in Product Backlog happened in only 8 cases of which 4 cases also reported a significant impact on the Product Vision, 2 reported a slight impact and the other 2 reported no impact on Product Vision.

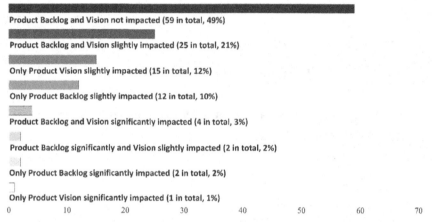

Fig. 7. The Covid-19 pandemic's impact on product backlog and vision

Figure 8 shows the change in stakeholders' involvement during the Covid-19 pandemic. In over half the cases the involvement remained the same. Stakeholders involvement increased in 25 cases (20.8%). Only in 6 of the 25 cases, where the stakeholders' involvement increased, did Product Vision and Product Backlog stay the same. On the other hand in 10 of 25 cases, where stakeholders' involvement increased, both the Product Vision and Backlog were impacted. The stakeholder's involvement decreased in 20 cases (16.7%). In 7 of these cases no impact on Product Backlog or Product Vision was reported. In 5 of these 20 cases both the Product Vision and Backlog were impacted.

The release frequencies of surveyed ASDT have been shown in Fig. 9. There is no dominant release frequency. Almost three quarters of the teams release at least every month. Almost half of the ASDT is releasing every 2 weeks or more often. Only 16% of the respondents are releasing every quarter.

Figure 10 shows the change in release frequency during the Covid-19 pandemic. Only 8 (6.6%) of respondents, state that they started releasing more frequently during the Covid-19 pandemic. In all of these cases the Product Backlog was changed, though only slightly in all cases but one (in which it changed significantly). In 6 of these cases the Product Vision changed slightly, only in 2 did it remained the same. Moreover 4 of these 8 cases where the release frequency was increased report that the Definition of Done (DoD) was made more liberal, in 3 cases it didn't change and in the last case the

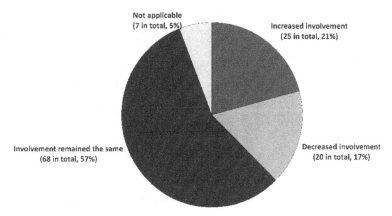

Fig. 8. Change in stakeholders' involvement during the Covid-19 pandemic

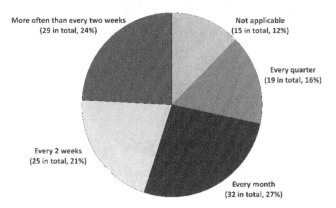

Fig. 9. Release frequencies of surveyed ASDT

team didn't have a DoD. Only 3 respondents report that during the Covid-19 pandemic they are releasing less frequently. In 2 of these cases they used to release more often than every 2 weeks and the third team was releasing every 2 weeks. In all of these 3 teams the DoD was not changed.

In Fig. 11 changes in DoD have been presented. The DoD was changed in only 14 cases. In 6 of these cases it became more liberal and was accompanied by a change in either Product Vision or Product Backlog. The DoD became more strict in 8 teams. All of these 8 teams didn't work fully remotely before the Covid-19 pandemic and changed to fully remote work. As many as 18 teams don't have a DoD, all of these teams except one use the Scrum framework.

In an open question respondents were asked what was the best change introduced in their work because of the Covid-19 pandemic. This question was not obligatory, consequently only 42 meaningful answers were gathered. Most of the respondents (30 from 42) indicated the introduction or maturing of remote work as the best change. From the rest 5 respondents see an increase in communication as the biggest positive

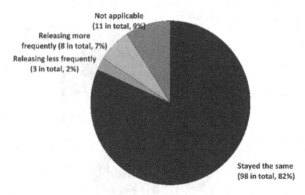

Fig. 10. Change in release frequencies because of the Covid-19 pandemic

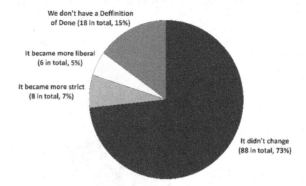

Fig. 11. Change in definition of done because of the Covid-19 pandemic

and another 4 indicated increased productivity. The remaining mentioned positive aspects were: reduced number of meetings, increased accountability, more automatization and more pair programing. According to the gathered answers, the communication increase was caused by moving all communication to online tools. Therefore, everyone had access to every discussion, while before people were omitted because they were remote at that moment or just not in the room where the discussion took place.

4.3   Metrics and Tools Used by ASDT

In the survey participants were asked to mark metrics used by their team and add any missing metrics. The total number of users for each metric has been presented in Fig. 12. The most popular metric was Velocity, with over half of the teams using it. The next most popular metrics were Quality, Work in Progress, Sprint Goal success Rate and Value Delivered. Only 13 out of 120 respondents did not report using any metrics, 8 of these 13 work in pure Scrum, 2 work in SAFe, 2 in pure Kanban and the last one uses both Scrum and Kanban.

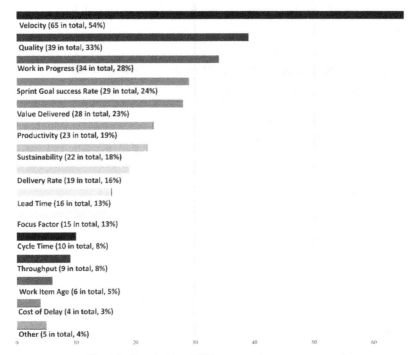

Fig. 12. Popularity of different metrics among respondents

In the next question respondents were asked if their team introduced any new metrics because of Covid-19. Only 5 participants reported that a new metric was introduced. These new metrics were:

- Focus Factor;
- Vanity Metrics;
- Daily resolved defects per team member;
- Skills gained and shared with the team;
- Weekly work hours reporting instead of monthly reporting.

The next question investigated what collaboration tools are used by the team. The answers have been presented in Fig. 13. Every ASDT uses at least one collaboration tool. The most popular tool is Jira, a task management tool, used by 77.5% of respondents. The second most popular tool is Confluence, a knowledge management tool closely integrated with Jira. The most popular communication tool is Teams (50%) with Slack (42.5%) being a close second. Another common tool is GitHub with its alternative GitLab behind it. These tools also have simple task and knowledge management functionalities in their primary feature of being a code repository. Next is less popular tool Azure DevOps which is both a task and knowledge management tool and a code repository. Later with 18 users there is Trello, a simple task management tool, and online whiteboard like

Miro and Conceptboard. The last of the commonly used tools is Mural, also an online whiteboard.

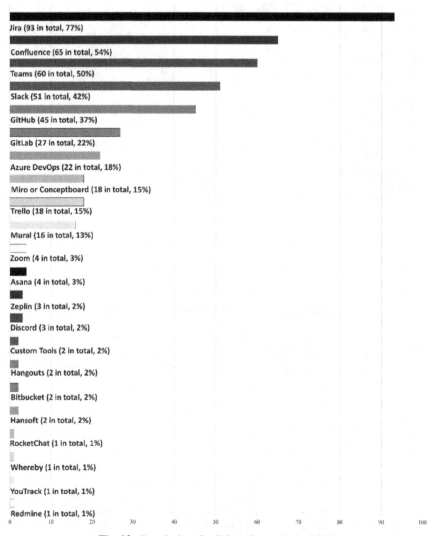

Fig. 13. Popularity of collaboration tools in ASDT

5   Discussion and Future Work

Obtained results may indicate that the ASDT were able to adapt to the new circumstances caused by Covid-19. The most common change that was observed is a shift towards remote work for almost all of the ASDT (85% of teams currently work fully remotely in response to Covid-19). The cause of such good adaptation can be found in the tools already used by the teams. The popularity of distributed teams, that have a similar characteristic to remote teams, resulted in the adaptation of commonly used tools and the development of features supporting work in distributed teams. Consequently, non-distributed non-local teams were already using tools supporting remote work prior to the Covid-19 pandemic. This minimalized the need to implement new tools or practices while transitioning to full remote work.

The popularity of different tools among ASDT indicates a significant need for more advanced knowledge sharing. The most popular tool was Jira, a task management tool. However, the second most popular tool was Confluence, used by over a half of the surveyed ASDT. This knowledge management tool, closely integrated with Jira, provides more advanced means of communication, supplementing simpler forms of direct messages and e-mails. Knowledge management functionalities can also be found in other tools popular among the participants including GitHub, GitLab and Azure DevOps. All of these tools offer multiple features including code repository, task management and knowledge management. Therefore, they can be used for different purposes in the ASDT. Exact use of these tools was not addressed in this study and should be further investigated in the future. Investigating which exact functionalities are used by the ASDT should allow for a better understanding of what ASDT need from their tools and provide input for their further improvement.

The disturbance in the business sphere of ASDT was surprisingly low. Almost half of the teams didn't change either their Product Backlog or their Product Vision. The release frequency and the DoD in most cases remained unchanged. The few ASDT that changed the workflow probably did so to address changes to the Product Backlog and Vision that were also reported by those teams. Despite the turbulence caused by the transition to remote work, most of the ASDT continued to deliver at the same rate and with the same quality. This may indicate that most of the products developed by the ASDT were not impacted by the pandemic and that the ASDT were able to transition to remote work without significant interference to their work. The lack of change in the business aspect of ASDT work can have multiple causes. The simplest one is a lack of influence of the pandemic on the products. The other reason may be the insufficient maturity of the Agile mindset among project managers and stakeholders. It is possible that ASDT have adapted the Agile frameworks but the formal restrictions don't allow for or don't require the ASDT to adjust the Product Vision or Backlog. Therefore ASDT are developing a product accordingly to the pre-pandemic circumstances. Investigating the cause for such small changes is a matter for future research.

Most of the ASDT didn't introduce any new metrics during the Covid-19 pandemic. Most of the teams used at least one metric, therefore they should be aware of the value metrics bring to the team and transparency of work. The lack of new metric introduction can be related to rather small changes in the organization of ASDT work during the pandemic. The other explanation can be a lack of understanding and recognition of metric usefulness in ASDT. Many of the respondents answered in the open question about why they use such metrics that they were chosen by the management or the organization, not by the ASDT themself. Few of the answers suggested a deeper understanding of the motivation behind the usage of metrics. The ASDT understanding and appreciation of metrics in software development should be further investigated in a future study.

6   Conclusion

The main objective of our research was to identify the impact of the Covid-19 pandemic and a rapid transition to remote work on the ASDT. Survey results prove the work of ASDT was not significantly impacted in most cases. The ASDT were able to transition to remote work without much turbulence. This smooth transition was possible due to the popularity of distributed and remote ASDT prior to the Covid-19 pandemic. Only 9% of all surveyed teams didn't work in a distributed team and didn't work remotely at all. The prevalence of distributed teams and remote work resulted in the popularity of online tools supporting it. Even the non-distributed, non-remote teams were already using online tools which support distributed teams. Therefore, the need to implement new tools in the ASDT was limited, which led to an easier transition to fully remote work. Accordingly, as not much was changed in the organization of ASDT work, most of the teams didn't feel a need to introduce new metrics. The business sphere of ASDT work was also not significantly impacted. In few cases the surveyed ASDT responded to the changes in Product Backlog and Vision by accelerating the work flow. They increased the release frequency and in a few cases lowered the overall quality for a short term speed gain. Such behavior could help with a quicker response to new market opportunities.

Results indicate that the transition to remote work didn't disrupt ASDT' communication. Rather, respondents state that fully remote work reduced the amount of unnecessary meetings, which were reducing their productivity. Moreover, fully remote work prevented the exclusion of remote or distributed team members from on-site, in-person discussions and meeting. Therefore, fully remote work improved communication in teams that were distributed and non-remote before the pandemic by moving all communication to online tools. Co-located team members couldn't exclude their distributed colleagues by discussing issues in person.

Over 89% of surveyed ASDT use at least one metric. The use of metrics supports their software development process and allows for making data driven decisions. The most commonly used metric was Velocity, used by over half of surveyed ASDT.

Results show that every surveyed ASDT uses at least one cooperation tool, including all the non-distributed, co-located ASDT. The most commonly used tool is Jira, a task management tool, and Confluence, a knowledge management tool. This shows that a need for more advanced cooperation tools is well known among ASDT and they are using them even when working in a non-distributed, co-located environment. The use of online tools allows each of the 120 respondents to have easy access to the Product Backlog while working remotely.

References

1. State of Agile Homepage. https://www.stateofagile.com/. Accessed 30 Sept 2020
2. Sharma, S., Hasteer, N.: A comprehensive study on state of Scrum development In: 2016 International Conference on Computing, Communication and Automation (ICCCA), pp. 867–872. IEEE, Noida (2016)

3. Berczuk, S.: Back to basics: the role of agile principles in success with an distributed scrum team. In Agile 2007, pp. 382–388. IEEE (2007)
4. Paasivaara, M., Behm, B., Lassenius, C., Hallikainen, M.: Large-scale agile transformation at Ericsson: a case study. Empirical Softw. Eng. **23**(5), 2550–2596 (2018). https://doi.org/10.1007/s10664-017-9555-8
5. Nevo, S., Chengalur-Smith, I.: Enhancing the performance of software development virtual teams through the use of agile methods: a pilot study. In: 2011 44th Hawaii International Conference on System Sciences, pp. 1–10. IEEE (2011)
6. Robinson, P.T.: Communication network in an agile distributed software development team. In: 2019 ACM/IEEE 14th International Conference on Global Software Engineering, pp. 100–104. IEEE (2019)
7. Stray, V., Moe, N.B., Noroozi, M.: Slack me if you can! using enterprise social networking tools in virtual agile teams. In: 2019 ACM/IEEE 14th International Conference on Global Software Engineering, pp. 111–121. IEEE (2019)
8. Vax, M., Michaud, S.: Distributed agile: growing a practice together. In: Agile 2008, pp. 310–314. IEEE (2008)
9. Cristal, M., Wildt, D., Prikladnicki, R.: Usage of Scrum practices within a global company. In 2008 IEEE International Conference on Global Software Engineering, pp. 222–226. IEEE (2008)
10. Downey, S., Sutherland, J.: Scrum metrics for hyperproductive teams: how they fly like fighter aircraft. In 2013 46th Hawaii International Conference on System Sciences, pp. 4870–4878. IEEE (2013)
11. Bartik, A.W., Bertrand, M., Cullen, Z., Glaeser, E.L., Luca, M., Stanton, C.: The impact of COVID-19 on small business outcomes and expectations. Proc. Natl. Acad. Sci. **117**(30), 17656–17666 (2020)
12. McKibbin, W.J., Roshen, F.: The Global Macroeconomic Impacts of COVID-19: Seven Scenarios. CAMA Working Paper No. 19/2020 (2020)
13. Raišienė, A.G., Rapuano, V., Varkulevičiūtė, K., Stachová, K.: Working from home—who is happy? A survey of Lithuania's employees during the covid-19 quarantine period. Sustainability **12**(13), 5332 (2020)
14. Mancl, D., Fraser, Steven D.: COVID-19's influence on the future of agile. In: Paasivaara, M., Kruchten, P. (eds.) XP 2020. LNBIP, vol. 396, pp. 309–316. Springer, Cham (2020). https://doi.org/10.1007/978-3-030-58858-8_32
15. Schwaber, K.: SCRUM development process. In: Sutherland, J., Casanave, C., Miller, J., Patel, P., Hollowell, G. (eds.) Business Object Design and Implementation. Springer, London (1997). https://doi.org/10.1007/978-1-4471-0947-1_11
16. Hartmann, D., Dymond, R.: Appropriate agile measurement: using metrics and diagnostics to deliver business value. In: AGILE 2006, pp. 126–131. IEEE (2006)
17. Scrumban, L.C., et al.: Essays on Kanban Systems for Lean Software Development. A Division of Modus Cooperandi. Inc.–Seattle, USA (2008)
18. Anderson, D.J.: Kanban: Successful Evolutionary Change for Your Technology Business. Blue Hole Press (2010)
19. Ohno, T.: Toyota Production System: Beyond Large-Scale Production. Productivity Press, Abingdon (1988)
20. Kupiainen, E., Mäntylä, M.V., Itkonen, J.: Using metrics in agile and lean software development–a systematic literature review of industrial studies. Inf. Softw. Technol. **62**, 143–163 (2015)
21. Kurnia, R., Ridi F., Sunu W.: Software metrics classification for agile scrum process: a literature review. In: 2018 International Seminar on Research of Information Technology and Intelligent Systems, pp. 174–179. IEEE (2018)

The Sars-Cov-2 Pandemic and Agile Methodologies in Software Development: A Multiple Case Study in Germany

Michael Neumann[✉], Yevgen Bogdanov, Martin Lier, and Lars Baumann

Hochschule Hannover, Ricklinger Stadtweg 120, 30459 Hannover, Germany
michael.neumann@hs-hannover.de

Abstract. In recent years, agile methodologies have been established in software development. Today, many companies use agile or hybrid approaches in software development projects. The Sars-Cov-2 pandemic has led to a paradigm shift in the way people work in Germany. While it was customary for German software development teams to work co-located before the pandemic, teams have been working on a distributed remote basis since March 2020. Studies show that distributed work impacts the performance of agile software development teams. To examine the effects of the Sars-Cov-2 pandemic on agile software development in Germany, we planned, carried out, and evaluated a multiple case study with three cases. The results show that the majority of teams did not experience any loss in performance. We present some problems and challenges and derive specific suggestions for software development practice from the results of the study.

Keywords: Agile methodologies · Agile software development · Distributed agile software development · Influencing factors · Success factors · Case study

1 Introduction

The Sars-Cov-2 pandemic has led to significant changes in social life. In Germany, many companies with a software development context have enabled their staff to work remotely since March 2020. Since then, more and more software development teams have been working distributed. Many of these teams were not accustomed to distributed work and were confronted with new challenges.

Agile and lean methodologies such as Scrum, Extreme Programming, or Kanban are established approaches in software development and are correspondingly widespread (cf. [22, 42]). Communication and collaboration are important aspects of agile approaches. The agile manifesto describes values and principles of interaction and collaboration and assigns them great importance. Consequently, these aspects are taken into account in the guidelines of agile methodologies (e.g. [35]). Many studies confirm the importance of various facets of communication and collaboration by identifying and describing them as success factors:

- Communication (cf. [24, 27, 33, 39])

© Springer Nature Switzerland AG 2021
A. Przybyłek et al. (Eds.): LASD 2021, LNBIP 408, pp. 40–58, 2021.
https://doi.org/10.1007/978-3-030-67084-9_3

- Integration and collaboration with the customer (cf. [5–7, 16, 19, 27, 29, 34, 39, 40])
- Aspects of teamwork (cf. [13, 16, 25])
- The development of self-organization (cf. [7, 39])

Various studies show the connection between distributed software development and project success. Nguyen et al. examine in [28] the possibilities of remote communication and the associated effects on cooperation within the software development team. Karolak describes in [21] the virtual leadership of software development teams and compliance with quality standards as a challenge of distributed software development. In addition, other aspects are presented in literature, such as knowledge management in distributed teams [14] or ensuring high-quality communication [28]. Therefore, we initially assumed that an ad hoc switch to distributed remote work would negatively impact the performance of agile software development teams. After discussions with practitioners, however, a different picture emerged. The perception is that the teams are at least as performant as before. Despite the potential increase in motivation, due to the nature of this pandemic, or the potential increase in working hours, we identify the following new challenges for the teams:

- Lack of Face-to-face communication
- Analog tools (such as physical task boards) must be digitized
- Integration and involvement of stakeholders is more difficult
- Set up effective training and coaching on agile methods
- Training of new employees is more difficult

Currently, no known publication addresses this topic of how the performance of agile software development teams is affected by challenges which arised from the Sars-Cov-2 pandemic. Furthermore, previous research is limited to the context of global software development in which software development teams are usually considered at companies in the outsourcing environment and are often used for distributed software development. However, this context cannot be immediately compared with the situation in Germany since the Sars-Cov-2 pandemic arose. Agile software development teams were usually co-located and are not used to working in a distributed environment. These agile teams had to transition quickly and effectively to the new distributed way of work. Also, studies on remote work in agile software development were published in the past, but the effects of the Sars-Cov-2 pandemic were not taken into account. One of these effects is the adhoc switch to the remote activity of entire teams, departments or even companies. It should also be noted that the changes in the remote activity of permanent nature of at least several months and at times no possibility to work as a co-located team existed. From our perspective, the conditions caused by the Sars-Cov-2 are therefore not comparable with the context of investigations at that time.

The paper is organized as follows: First, in Sect. 2, we summarize related work in distributed and remote agile software development as well as agile software development teams during the Sars-Cov-2 pandemic. In Sect. 3, we describe the research questions and methodology of the multiple case study. The results are presented in Sect. 4. Section 5 covers the limitations of this paper. Finally, in Sect. 6, we summarize the results and give an outlook on our future work.

2   Related Work

2.1   Agile Methodologies in Distributed and Remote Working Software Development

The basis of agile methodologies in software development is the agile manifesto [2]. The core idea behind creating the agile manifesto, encompassing the four value statements and twelve principles, was to offer a uniform philosophy for agile methodologies in software development.

According to Abrahamson et al. in [1], agile approaches are characterized by the following properties:

- Incremental: Small software releases in short cycles
- Cooperative: Customers and software developers constantly work closely together
- Straightforward: The process is easy to learn, adapt and well documented
- Adaptable: It is possible to make changes to the product at the last possible moment

Beck describes agile approaches as efficient, flexible, low-risk and predictable [3]. Agile methodologies are iteratively structured and aim for fast response times during the project period [17].

In this article's context, a distinction must first be made between distributed and remote working agile software development. In distributed software development, the connection to global software development (GSD) is usually considered. At GSD, software development projects are implemented in international cooperation [11, 18, 23]. This international cooperation arises, for example, through the outsourcing of software development projects [9]. Software is usually not implemented in the same geographical region in which it is commissioned or used. This is often described as distributed software development, in which the members of software development teams work at several locations [41]. Various challenges and problems accompany distributed software development. Effective communication is necessary for successful software development [31]. Shah et al. address how this can be ensured in global software development [38]. Nguyen et al. explain in [28] the possibilities of remote communication and the associated effects on cooperation within the software development team. Ebert and Neve [15] point out, among other things, that different languages or language levels can represent a potential barrier in communication. Furthermore, the challenges of virtual leadership of software development teams, as well as ensuring a working development environment and compliance with quality standards, should be mentioned [21]. Dingsoyr and Smite [14] also present the special features of knowledge management in global software development and recommend which approach should be used in this context.

Due to the Sars-Cov-2 pandemic, a fundamental change in working has been observed in Germany. While before the pandemic, work was mostly co-located in the software development environment, and the home office, for example, was a rarity, this has been different since the beginning of Sars-Cov-2. Many companies have switched to enabling or allowing parts of the workforce to work remotely. Software development teams are also increasingly affected by this. This change has led to a distributed activity of software development teams in Germany as well. Incidentally, this also prevails when parts of the team are in the office. The reason for this is that it has effects on the entire team if individual team members work from other locations. These effects can manifest themselves, for example, through the virtualization of the implementation of agile practices (such as daily meetings) or artifacts (e.g., Kanban boards).

It seems problematic to distinguish the differences between distributed teams and remote work clearly. Some of the challenges of distributed agile software development in the context of GSD, such as language barriers or time zone differences, are certainly negligible in the current situation in Germany. Other aspects, for example, concerning virtual communication and collaboration, are relevant. In literature, remote work is also described with the characteristics of distributed teams [12]. However, the distributed activity in software development projects poses a particular challenge when using agile approaches. The agile manifesto and various guidelines of agile approaches emphasize the importance of communication and collaboration. However, as explained above, this is more difficult with distributed activity. Various studies confirm that communication influences the success of agile methodologies in software development. Sinha et al. represent the effectiveness of communication as a success factor in [39] in global software development. The effectiveness of the communication channels is particularly relevant at different development locations. The authors Borrego et al. also point to the importance of communication in agile global software development [4]. In [10], Chow and Cao name another aspect of communication: direct "face to face" communication in daily meetings. Considering the communication and interaction of agile software development teams, the choice and use of the right agile practices are also important. In [37], Senapathi and Srinivasan relate this choice of agile practices, among other things, to the continuous optimization of the development team's approach. The authors state that the in-depth use of agile practices is important. This in-depth use is described by the example of Kanban practices and, according to the authors, leads to positive effects on performance. They also describe the individualization of agile practices in different teams (also within a company). This individualization can arise for different reasons, for example the experience of the teams with an agile approach. In their qualitative study, Senapathi and Drury-Grogan in [36] confirm the importance of the right choice of agile practices as a success factor.

2.2 Agile Software Development Teams During the Sars-Cov-2 Pandemic

The challenges of agile software development teams triggered by the Sars-Cov-2 pandemic are the subject of initial publications in the literature. Various papers deal with how agile software development teams have mastered the initial challenges of distributed remote work [8, 26]. For this purpose, the teams used Microsoft Teams, Slack or zoom to maintain communication and digital whiteboards were used to ensure continued collaboration within the team.

A concrete procedure in the distributed remote activity describes da Camara et al. in [8] using the example of a Brazilian start-up that maintains agile software development teams. In Action Research, the authors present 23 specific measures intended to support agile software development teams in meeting the challenges of distributed remote work. The measures are both of an organizational nature (e.g., the provision of necessary hardware and software) and the process-specific nature of the agile approach (e.g., selecting a tool for the sprint retrospective or the introduction of code reviews). They find that the measures taken have had positive effects. For this, they state the source code's quality, the sharing of knowledge within the team, or the understanding of the project requirements.

Poth et al. present in [30] a digital service approach to increase the distribution of knowledge in companies and to increase team autonomy. To do this, use the Self-Service Kit (SSK), which the authors define as a combination of various learning and training methods. The SSK activates teams and departments at any hierarchical level to share or build up existing and new knowledge, despite distributed organizational forms. Poth et al. also refer in [30] to the effects of the Sars-Cov-2 pandemic on the intensified relevance of the SSK, as it supports the teams in meeting the challenges of distributed remote work. They also point out that social interaction is still preferred, while remote activity is accepted as a form of work.

Also, other effects on work organization in companies and administrations are described (e.g. in [20, 26]). Some companies are already starting to let their employees decide where they want to work and live remotely. Reference is also made to the risks of remote work, for example, in the context of occupational safety. In addition to these aspects of work organization, reference is also made because of the effects on the recruiting of new employees and the subsequent integration of these new team members into the agile software development teams require optimized strategies for virtual induction.

The literature presents concrete results in the context of the new challenges of agile software development teams due to the Sars-Cov-2 pandemic. Even if these present valuable insights, the context or the objective of the individual publications is different.

3 Research Design

3.1 Research Approach

We use the multiple case study approach. For the planning, preparation, implementation, and evaluation of the study, we use the guidelines from Runeson and Höst from [32]. The authors describe how the case study method is suitable for gaining a deep understanding of the "phenomena under study" [32]. Additionally, they state that case studies are suitable for field research. Runeson and Höst describe in [32] various characteristics of case studies. We choose the exploratory case study for the present paper because existing literature helps formulate the research object concretely and narrow down the context. According to Runeson and Höst in [32], the exploratory case study is suitable for this as it enables "... finding out what is happening, seeking new insights and generating ideas and hypotheses for new research..." [32].

The research context is limited to organizations in Germany that use agile or hybrid methodologies in distributed software development and already used these approaches before the Sars-Cov-2 pandemic. We justify this contextual limitation as follows. In Germany, we can understand the change from co-located work to distributed software development due to the Sars-Cov-2 pandemic. The cases included in the study are selected accordingly. Secondly, we can precisely describe the time period, i.e. the second half of March 2020, in which the teams started their transition to distributed software development. Further contextual restrictions, e.g., to industries, are not considered relevant. This is because the challenges mentioned above are classified as relevant regardless of industry. For this study, we choose a holistic approach. The multiple case study comprises three cases, each representing a company in Germany. In these cases, we understand the agile-organized software development team as a unit of analysis. This approach enables us to collect maximum data from the agile software development of a company (case). We assume that software development in the respective cases is similarly impacted by the ad hoc switching to distributed agile software development. The three cases are different companies that operate in different industries and markets. They also differ in size and corporate culture. Case 1 is a group that, among other things, operates the end-customer business in e-commerce. Case 2 is a medium-sized company that offers various software products (including Enterprise Resource Planning applications) and consulting for business customers. Case 3 is also a medium-sized company that develops websites and mobile apps for its business customers.

3.2 Research Questions

As explained in Sect. 2, various studies show that the use of agile methodologies in distributed software development entails multiple challenges, for example, in communication or collaboration. We also describe above that some of these aspects influence the success of agile methodologies. Due to the distributed activity in software development, triggered by the Sars-Cov-2 pandemic in Germany, we assume that the performance of agile teams has changed. Furthermore, we want to investigate what effects can be observed on certain success factors in agile software development. We focus this work on the communication of the team, the integration and cooperation with the customer, and the adaptation (and selection) of agile practices. The research questions specify the objectives of this paper as follows:

- **RQ 1:** *Has the performance of agile software development teams changed since the beginning of the Sars-Cov-2 pandemic?*
  First of all, we seek verify whether the performance of agile software development teams has changed during the Sars-Cov-2 pandemic.
- **RQ 2:** *Were agile practices or roles adapted during the implementation of distributed software development? If so, which practices or roles are affected, and how have they been changed?*
  We assume that the use of agile practices has changed during the Sars-Cov-2 pandemic. This is justified by the empirical and self-optimizing approach in agile software development. In this context, we examine which practices have been adapted and how. Furthermore, we aim to identify and understand any newly introduced practices.
- **RQ 3:** *How has communication and collaboration in agile software development teams changed due to the pandemic?*
  Communication and collaboration are subject to changes due to the shift to distributed software development. We aim to understand precisely how communication and collaboration in agile software development teams have changed and what effects this has on their success (e.g. in terms of productivity or customer satisfaction).
- **RQ 4:** *What has changed in terms of the collaboration and involvement of the customer/stakeholder?*

The integration and cooperation with the customer are emphasized in agile software development. Moving to distributed software development, it can be assumed that this aspect is also subject to changes.

3.3 Data Collection

In exploratory case studies, the qualitative data collection is common (cf. [32, 43, 44]). According to Runeson and Höst in [32], triangulation is of great importance in qualitative research approaches. The authors justify this in [32] because qualitative data is *"... broader and richer, but less precise than quantitative data."*. With the help of triangulation, the validity and reliability of the data should be improved.

In this paper, we use both direct methods, such as semi-structured interviews and the observation of agile practices, and independent analysis to collect data. As part of this analysis, we sift through, for example, project documentation and artifacts from the agile software development teams. We have developed interview guidelines for conducting the semi-structured interviews (see Appendix A). The interview guidelines include both open and closed questions. The interviews follow an identical organizational scheme. First, the interviewing researcher presents the goals and context of the study and explains how the interview data is processed. This is followed by the interview, in which general questions are asked first. For example, we inquire about the experience in software development projects, the current role in the team and the agile or hybrid methodologies currently in use. The content of the rest of the guideline is based on the research questions (see Sect. 3.2). These questions are, therefore, more specific to the objectives of this study. If further questions arise, these can be asked by the interviewer. If this occurs, these questions will be documented in the interview by the protocol officer. If possible, the interviews will be carried out by two researchers. While one researcher conducts the interview, the second researcher logs the interview. Further researchers are involved in the subsequent data evaluation to reduce potential bias and increase objectivity.

As stated above, we also collect data by observing agile practices. These can include, for example, meetings described in the literature, such as a sprint retrospective, but also adapted practices. The observations are documented with the help of a prepared protocol (see Appendix B). During the observations, we try to assume a passive role and trigger little to no interaction with the respective team. In conjunction with these direct methods of data collection, we also examine various artifacts of the teams. In addition to product backlogs or Kanban boards, this can include project documentation such as team performance evaluations. The data was collected during August and the first half of September 2020. We assume that the performance (RQ1) and the factors influencing adapted agile elements (roles, artifacts, practices), as well as communication and collaboration in the team and with the customer (RQ 2 to RQ 4), were not stable in the first weeks after the Sars-Cov-2 pandemic hit Germany in March 2020. The agile teams had the chance to react to the (organizational) changes which occurred after the switch to distributed software development. We expect that in the meantime the teams adapted their agile approach to this new situation and that the performance is now in a stable state.

3.4 Data Analysis

We have structured the analysis of the data based on the guidelines by Runeson and Höst from [32]. We extracted the data from the interview and observation protocols and structured it in tabular form and defined different areas of the evaluation tables using the research questions as a structural basis. Next, we assigned data from the interviews to the questions in the interview guidelines and also structured the protocols of the observations accordingly. This enabled a structured overview of the data. Based on these tables, we initially looked for similarities in the data. Each researcher did so individually and commented on the mentioned similarities accordingly. If we identified at least three data points (interviews, observations, or document reviews) with similar or identical statements, these data were marked appropriately and revised by at least one other researcher. If the quality could be assured, the data were then extracted. We used Miro to visually represent our extracted data, employing virtual whiteboards, which met the needs of our own distributed activities. We verified the extracted data on our virtual whiteboard together. Finally, we analyzed the extracted data on the structural basis of the research questions.

4   Results

4.1   Overview of the Results

Scrum is the dominant agile methodology in practice (see Sect. 1). This is also confirmed in this multiple case study (see Table 1). Cases 2 and 3 work with Scrum. Case 1 uses a mixed approach with Kanban.

Table 1. Agile methodologies used per case

Case	Agile methodologies in use
Case 1	Kanban, mixed approach of Scrum and Kanban
Case 2	Scrum
Case 3	Scrum

To get a more detailed insight into the application of the used agile methodologies in the cases, we asked the interview partner which agile practices and artifacts were being used. The results are summarized in Table 2. The assignment of the agile practice or the artifact to the respective case is shown with an x. It has been shown here that practices that we know, for example, from Scrum (such as the daily stand-up, planning, or even the retrospective) take place in all three cases. We were also able to identify pair programming in all three cases. The use of the Kanban board correlates with the use of the hybrid process model in Case 1 (see Table 1). We only identified the practice of Continuous Deployment in Case 1.

It should be noted at this point that the artifact practices can also differ in the way they are acted out within a case depending on the team. This applies to both the use and

Table 2. Identified agile practices and artifacts

Agile artifacts and practices in use	Case 1	Case 2	Case 3
Code review	x	x	x
Continuous deployment	x		
Coding standards	x	x	x
Daily stand up	x	x	x
Definition of done	x	x	x
Estimation	x		
Kanban board	x		
Pair programming	x	x	x
Planning meeting	x	x	x
Product backlog	x	x	x
Retrospective meeting	x	x	x
Review meeting	x	x	x
Sprint backlog	x	x	x
Test driven development	x	x	

the specific application. The specific application is not decisive at this point. Rather, we wanted an overview of which practices and artifacts are used in the respective teams. This helps us get an impression of how the respective team operates.

Interestingly, certain agile practices in distributed work have a double-edged character. An example of this is pair programming, which is perceived as challenging by the software developer in the distributed activity. On the other hand, it helps to maintain the exchange between team members, has a positive influence on the product quality and ensures an exchange of information. The team members also benefit from internal transfer of experience and skills. The requirements for the data collection, e.g. data anonymization, were coordinated with the contact persons at the respective companies (cases). The perception of different aspects of this study, such as communication, can vary depending on the role and point of view. Therefore, it is sensible to consider all roles directly involved in the agile approach in the survey. This requirement was successfully taken into account. An overview of the profiles of the interviewed persons is shown in Table 3.

We also deliberately omitted the specific terms relating to experience, such as junior or senior software developer. These are not defined uniformly across different companies. Instead, we decided to consider the experience in software development projects and with methodologies in software development.

We conducted a total of 22 interviews, eight each for Cases 1 and 2, and six more for Case 3. As explained in Sect. 3, we also collected data through observations of agile practices. We observed various teams performing 18 agile practices. A summary of these observations is presented in Table 4.

Table 3. Interview profiles

Case	ID	Actual role	Years of experience in SD	Years of experience in ASD[a]	Team
Case 1	P01	Product owner	21	9	C1-A
Case 1	P02	Product owner	15	10	C1-B
Case 1	P03	Software developer	10	8	C1-C
Case 1	P04	Software developer	6	6	C1-D
Case 1	P05	Software developer	10	10	C1-E
Case 1	P06	Agile coach	24	5	C1-F
Case 1	P07	Product owner	12	9	C1-G
Case 1	P08	Agile coach	17	15	C1-B
Case 2	P09	Software developer	4	4	C2-A
Case 2	P10	Scrum master	15	6	C2-B
Case 2	P11	Software developer	7	3	C2-B
Case 2	P12	Software developer	6	4	C2-C
Case 2	P13	Software developer	15	10	C2-D
Case 2	P14	Scrum master	22	22	C2-E
Case 2	P15	Software developer	23	7	C2-B
Case 2	P16	Product owner	28	3	C2-F
Case 3	P17	Product owner	1	1	C3-A
Case 3	P18	Software developer	3	3	C3-A
Case 3	P19	Software developer	10	8	C3-A
Case 3	P20	Scrum master	4	4	C3-A
Case 3	P21	Software developer	7	2	C3-B
Case 3	P22	Scrum master	4	2	C3-B

[a]Years of experience in agile software development is part of total experience in software development

It is not surprising that Scrum practices were most represented in the observations, as Scrum is the dominating approach in this case study (it is used in two out of three cases; see Table 1). In addition to these common practices, such as daily Scrum, sprint reviews or planning and refinements, we also observed other practices such as a coding session and an estimation meeting.

4.2 Answering the Research Questions

In the following, we will present the concrete results of this multiple case study. For readability, we structure the results based on the research questions from Sect. 3.2:

Table 4. Observed agile practices

Case	ID	Agile practice	Team
Case 1	E01	Team time out	C1-A
Case 1	E02	Daily stand up	C2-B
Case 1	E03	Estimation	C2-B
Case 1	E04	Coding session	C2-B
Case 1	E05	Retrospective	C2-B
Case 2	E06	Daily scrum	C2-A
Case 2	E07	Sprint retrospective	C2-B
Case 2	E08	Sprint review	C2-B
Case 2	E09	Sprint planning	C2-B
Case 2	E10	Sprint planning	C2-A
Case 3	E11	Sprint review	C3-A
Case 3	E12	Sprint retrospective	C3-A
Case 3	E13	Sprint planning	C3-A
Case 3	E14	Refinement	C3-A
Case 3	E15	Daily scrum	C3-B
Case 3	E16	Refinement	C3-B
Case 3	E17	Daily scrum	C3-B
Case 3	E18	Sprint review	C3-B

RQ 1: *Has the performance of agile software development teams changed since the beginning of the Sars-Cov-2 pandemic?*

The efficiency and performance of agile teams in the three cases of this multiple case study have not decreased. In the interviews, ten people said that the performance had increased (P03, P05, P06, P07, P11, P14, P15, P17, P18, P22). Twelve people stated that the performance was similar (P01, P02, P04, P08, P09, P10, P12, P13, P16, P19, P20, P21). These statements were verified and confirmed by viewing documentation such as Sprint Goal attainments in all three cases.

This finding is interesting insofar as we assumed that the challenges of switching to distributed agile software development would negatively affect performance. Hence, we investigated what the teams in this case study were doing to prevent these harmful effects. Based on the data collected, this can be narrowed down to three aspects:

1. Increased transparency of the process

In several interviews in all three cases, the conclusion could be made, that the agile approach became more transparent. This increased transparency is exemplified, e.g., by digitizing Kanban boards (P06), which visualizes the changes in the processing status of individual items ad hoc. Likewise, more people after the transition to distributed work have access to the board, increasing transparency even beyond the team. Another aspect that has increased transparency are the forms of communication (see also RQ3). Thanks to the digitized team-wide communication through tools, more team members receive questions and clarifications (P08). Additionally, communication is more factual and, therefore, the result is more precise (P04).

Agile approaches are usually based on empirical procedures. The Scrum Guide names transparency as one of the three pillars of empiricism [32]. Transparency enables the teams to examine their approach in a targeted manner (cf. Scrum Guide [32]; inspect) and to adapt or optimize based on the findings (cf. Scrum Guide [32]; adapt). In the agile approach, transparency is a decisive factor for the performance of the teams. Based on this case study, we assume that the increased transparency had a positive effect on the agile approach. The teams had the opportunity to optimize their procedures in a more targeted manner so that potentially negative aspects of the distributed activity were at least offset. This can also be seen in the interviews (including (P03, P04) and observation of meetings (E01), especially retrospectives (E12, E18).

2. Working time is used more efficiently

At the beginning of the study, we made the assumption that the members of the teams worked quantitatively more hours. The interviews did not confirm this assumption. Instead, the current working time is used more efficiently in terms of quality (P03, P05, P06, P07, P08, P09, P13, P14, P17, P18, P19). This can be attributed to fewer interruptions during work than was previously the case in everyday office life. The team members can better control potential disruptions by using the present status in the communication tool (such as MS Teams; see RQ 3). Also, it was pointed out in interviews that the team members are sensitized to disturbing a colleague with questions or problems. In addition to this aspect, there are effects on social exchange (see RQ 3). This effect can also be demonstrated in the respective meetings. Since the beginning of the pandemic, meetings have been more goal oriented, factual, and more efficient.

3. Optimized integration of the Product Owner

The interviews also showed that tool-supported and virtual communication improves the integration of the product owner in some teams (see RQ 4). It can be assumed that clear communication adds to this.

RQ 2: *Were agile practices or roles adapted during the implementation of distributed software development? If so, which practices or roles are affected, and how have they been changed?*

In all three cases, we can attribute specific changes in agile practices to the pandemic or the switch to distributed work. These adaptations relate, for example, to the digitization of practices and artifacts such as Kanban boards that were administered and used analogously before the pandemic (P01, P03, P06). Changes have occurred in agile practices that were previously carried out at the team level in a room. This affects, among other things, the Daily Stand Up Meeting (Daily Scrum), which is now carried out daily at the team level to increase the synchronization of the team (P21). Previously, only weekly meetings were held for the team. Other teams have introduced another meeting after the Daily Stand Up Meeting to promote the exchange within the team (C19DE-C1-P03). The procedure was (compulsorily) digitized in planning and retrospectives. While previously estimates were made using Planning Poker cards (P20, P22), this is now done in communication applications (such as MS Teams, Slack). To avoid bias in estimates, the Scrum Master or a team member verbally counts down from three. Only then are the estimates published in the Team Channel and visible to everyone (E03, E09, E13, E14, P11, P18). For retrospectives, some teams use tools such as Retrium (E12, E18). Other teams document the results of the retrospective in MS-Word (E07). The methodology of the retrospectives, however, has not changed due to the pandemic. The practice of pair programming has changed in terms of decreased frequency, sequence, and intensity (P02, P03, P06, P19). Some test persons also stated that the quality of pair programming has decreased (P04). This is justified in particular by the effects of the distributed activity, such as restrictions due to digital collaboration (P04).

Other interviewed persons noted that no or only marginal changes had been made to agile practices, artifacts, or roles (P01, P04, P09, P10, P12, P15, P16, P17, P18).

We examined whether changes to the agile approach can be specifically attributed to the pandemic from the perspective of the test persons. This was partly denied; it was pointed out much more than it is often the usual optimization when using the agile methodology (P05, P11, P13). Examples of this are the adaptation of the iteration length (P13). By anchoring the constant optimization of most of these approaches, the change of practices, artifacts, and roles in the agile approach is not uncommon. In this respect, the realization that only a few adjustments to practices, artifacts, or roles have been made is also quite interesting. However, investigating this was not the focus of this paper and we refer the reader to the description of further research (Sect. 6).

RQ 3: *How has communication and collaboration in agile software development teams changed due to the pandemic?*

First and foremost, it can be stated that communication has become more objective and efficient (P01, P06, P08, P12, P13, P19, P20, P21, P22). This means that social exchange and togetherness have decreased or have almost entirely ceased. Situations like getting a coffee with a team member, taking a break in the tea kitchen, or undertaking joint activities as a team (such as lunch) are not possible in distributed software development. Some teams in Case 1 and 2 have, therefore, created meetings which primarily serve the purpose of social interaction and exchange. While these meetings are still being held for Case 1, they have now been discontinued for Case 2. In various interviews in Case 1 and 2, although the positive intention of these discussions was acknowledged, the purpose was also questioned. Nevertheless, these social practices seem to be a way to maintain social interaction in teams and mitigate some negative influence of the distributed work. In the interviews, we also raised the question of whether there has been an increase, decrease or no change in the number of conflicts in the agile teams since the beginning of the distributed activity (Appendix 1). The majority of the test subjects in all three cases confirmed that the number of conflicts has not changed (P07, P08, P09, P10, P13, P14, P15, P16 to P22). Strategies similar to those used before the pandemic (and distributed activity) are used to resolve these conflicts (P01, P06, P22). Individual interviewed persons, however, point out that non-verbal communication should accelerate the resolution of conflicts (P01, P07). Concerning the collaboration of agile software development teams, the technical infrastructure, especially in software applications for communication, play an essential role. An overview of which tools are used in the respective cases is shown in Table 5. It should be noted that companies deal differently with the availability and acquisition of such applications. In Case 1, for example, there are clear guidelines about which tools may be used (P01 to P08). In Case 2, it was pointed out which tools are already licensed and available in the company (P12 to P14). Although there are no specific restrictions, the intent is to prevent the purchase and use of too many different tools. There are no specifications for Case 3 (P17 to P22).

Table 5. Used tools for collaboration and communication

Case	Tools
Case 1	Confluence, Jira, Microsoft Outlook (Mails), Microsoft Planner, Microsoft Teams, Miro, Slack, Threema, Visual Studio Code, WhatsApp, Zoom
Case 2	Yammer, Microsoft Outlook (Mails), Microsoft Teams, Microsoft Team Foundation Server, VoIP phones
Case 3	Confluence, Jira, Microsoft Outlook (Mails), Microsoft Teams, Slack, Zoom

Agile teams in all three cases examined use Microsoft Teams. Other common tools are Zoom and Slack. It must be taken into account which agile practices have a significant influence on the collaboration. For example, the agile practice of pair programming is used in all three cases (see Table 2). Pair programming is carried out in very different ways. Some pairs use screen sharing via MS Teams, Zoom, or other applications. Other agile teams also evaluated specific plug-ins for the development environment. This evaluation is still on-going. It can be seen here that the agile teams attach great importance to pair programming in all three cases.

RQ 4: *What has changed in terms of the collaboration and involvement of the customer/stakeholder?*

In this case study, we consider the involvement of customers and stakeholders in particular in the context of the role of the product owner. In the interviews, it was said several times that this involvement was intensified (P01, P03, P17, P20). This is justified by the introduction of communication tools. Through the use of these tools, the product owner appears more accessible and, through the communication in these tools, also has a direct channel to his team members (P05, P08, P10, P16). This is true despite the prevailing asynchronous communication. Interestingly, some interview partners also note that the product owners are less involved in their teams (P17, P20). This is justified, among other things, by better self-organization (P07).

5 Threats to Validity

Although we followed the guidelines according to Runeson and Höst (see Sect. 3), some threats to validity must be taken into consideration for the present study.

Construct Validity: According to Runeson and Hoest in [32] *"This aspect of validity reflect to what extent the operational measures that are studied really represent what the researcher have in mind..."*.

Here we see a threat related to our interview guidelines. The guideline contains many questions and may lead to the respondents answering shorter or less extensive answers as the interview lasts. This can lead to misinterpretations. We always made sure to ask questions in the interviews when uncertainties of understanding occurred. Besides, the interview guidelines are structured based on the research questions.

Internal Validity: The internal validity focuses on the aspects of the study design and on issues relating to data collection and evaluation. A significant challenge in qualitative research projects in particular is securing the chain of evidence [32]. We have, therefore, carried out structured data analysis. During the data collection, we also applied a four-eyes principle and worked in pairs. The procedure and the results of the data analysis were adopted by researchers from our team who did not experience any bias from participating in the data collection. We also triangulated in different dimensions. When selecting the cases, we initially paid attention to the differences in industry, size, and company structure. When collecting the data, we took various methods into account with semi-structured interviews, observations of agile practices, and reviewing project documents. We always took into account the origin (based on the survey method and the case) of the data to triangulate the respective extracted data and the findings derived from it.

External Validity: The external validity refers to the generalization of the study results. Only three cases could be considered in this study. This can be an external threat. From our point of view, however, we have implemented a systematic approach. Also, it must be taken into account that the three cases are active in different industries and are also not comparable in terms of other aspects such as size or company form. We have, therefore, tried to achieve the most heterogeneous setting of the cases to be able to examine the effects of the pandemic on different companies and agile software development teams. Nevertheless, we have already started the second iteration of data collection in which we are integrating three further cases to validate the results (see Sect. 6). We are planning a quantitative study to be able to verify the findings in an international context and we assume that a change to distributed agile software development has also occurred in other countries.

6 Conclusion and Future Work

Sars-Cov-2 pandemic has changed the way we work. This is evident in Germany, through the change from co-located work to distributed work in the home office. Distributed remote agile software development goes hand in hand with various challenges posed by this transition (see Sect. 2). Some of these challenges can be related to specific success factors (see Sect. 1). Based on this, we examined in a multiple case study to what extent the transition to distributed and remote work, due to the Sars-Cov-2 pandemic, affects the performance of agile software development teams and specific aspects of the agile approach (see Sect. 3.2). The multiple case study comprises three different cases. We collected the data qualitatively with semi-structured interviews, observations of agile practices, and viewing documentation (such as performance evaluations) (see Sect. 3).

In this paper, we present our qualitative multiple case study (see Sect. 3.1). We have found that the performance of agile software development teams has remained the same or even improved since March 2020 (see Sect. 4.2). We can justify this result with observations of increased transparency in the agile approach, qualitatively more efficient working hours, and the optimized integration and involvement of the product owner (see Sect. 4.2; RQ 1).

We also recorded various adjustments to agile practices, artifacts, and roles such as the necessary digitization of analog artifacts (Kanban boards) when switching to distributed activity. Other adjustments can be traced back to the expected striving for optimization in agile approaches (see Sect. 4.2; RQ 2). Concerning practices, artifacts, and roles that have not or only slightly adapted (see Sect. 4.2), further research is necessary. Specifically, we would like to address whether fundamental changes in the organization of agile teams, such as the switch to distributed software development, affect the team's motivation for the change of the process. This can also be influenced by other factors such as the maturity of agile methodologies in the respective team or company, or how long the team has been working together.

We also examined the impact of communication. We first analyzed which tools are used in the teams for communication and collaboration. Besides MS Teams, these are especially Zoom and Slack. We also looked at the effects of this digital communication and collaboration. Since the personal character of distributed communication through these tools is often left by the wayside, some teams have, for example, create new meetings for this specific purpose to continue to ensure personal exchange. We have also found that communication is more objective and transparent. There are fewer misunderstandings, which helps increase efficiency (see Sect. 4.2; RQ 3).

Lastly, we dealt with the involvement of and cooperation with the product owner. In many teams, this is has improved. One reason is that there is better accessibility in the asynchronous availability of the product owner in the respective communication tools (see Sect. 4.2; RQ 4).

We have already started the second iteration of this case study, in which we are integrating three more cases. The three new cases differ to those of this case study in industry, size, and structure. This second iteration aims, in addition to the aspects mentioned above, to validate the findings in other contexts (companies). We hope this will give us more information about the extent to which our results can be transferred to different organizations. In addition to the second iteration of this multiple case study, we plan to conduct a quantitative investigation to validate our findings. We are considering carrying out this quantitative study in other countries to validate the results in an international context.

Appendix 1

We have made the interview guideline used available in the Academic Cloud: Download Link (https://sync.academiccloud.de/index.php/s/gjfQ8AFDsEA77Rx).

Appendix 2

The observation protocol template is available in Academic Cloud: Download Link (https://sync.academiccloud.de/index.php/s/CPc0GKnsC0ZVV0v).

References

1. Abrahamsson, P., et al.: Agile software development methods: review and analysis, pp. 7–94 (2002)
2. Beck, K., et al.: Agile Manifesto. https://agilemanifesto.org/
3. Beck, K.: Extreme Programming Explained. Embrace Change. Addison-Wesley, Boston (2000)
4. Borrego, G., et al.: Towards a reduction in architectural knowledge vaporization during agile global software development. Inf. Softw. Technol. **112**, 68–82 (2019)
5. Bjarnason, E., et al.: A theory of distances in software engineering. Inf. Softw. Technol. **70**, 204–219 (2016)
6. Bermejo, P.H.d.S., et al.: Agile principles and achievement of success in software development. A quantitative study in brazilian organizations. Procedia Technol. **16**, 718–727 (2014)
7. Barzilay, O., Urquhart, C.: Understanding reuse of software examples. A case study of prejudice in a community of practice. Inf. Softw. Technol. **56**, 1613–1628 (2014)
8. Camara, R., Marinho, M.L., Sampaio, S., Cadete, S.: How do Agile Software Startups deal with uncertainties by Covid-19 pandemic? Int. J. Softw. Eng. Appl. (2020)
9. Conchúir, E.Ó., et al.: Global software development: where are the benefits? Commun. ACM **52**, 127 (2009)
10. Chow, T., Cao, D.-B.: A survey study of critical success factors in agile software projects. J. Syst. Softw. **81**, 961–971 (2008)
11. Chadee, D., Raman, R., Michailova, S.: Sources of competitiveness of offshore IT service providers in India: towards a conceptual framework. Compet. Chang. **15**, 196–220 (2011)
12. Deshpande, A., Barroca, L., Sharp, H., Gregory, P.: Remote working and collaboration in agile teams. In: International Conference on Information Systems (2016)
13. Drury-Grogan, M.L.: Performance on agile teams. Relating iteration objectives and critical decisions to project management success factors. Inf. Softw. Technol. **56**, 506–515 (2014)
14. Dingsoyr, T., Smite, D.: Managing knowledge in global software development projects. IT Prof. **16**, 22–29 (2014)
15. Ebert, C., de Neve, P.: Surviving global software development. IEEE Softw. **18**, 62–69 (2001)
16. Gren, L., Knauss, A., Stettina, C.J.: Non-technical individual skills are weakly connected to the maturity of agile practices. Inf. Softw. Technol. **99**, 11–20 (2018)
17. Flora, H.K., Chande, S.v.: A systematic study on agile software development methodologies and practices. Int. J. Comput. Sci. Inf. Technol. **5**, 3627–3637 (2014)
18. Herbsleb, J.D., Moitra, D.: Global software development. IEEE Softw. **18**, 16–20 (2001)
19. Hoda, R., Noble, J., Marshall, S.: The impact of inadequate customer collaboration on self-organizing Agile teams. Inf. Softw. Technol. **53**, 521–534 (2011)

20. Janssen, M., van der Vorrt, H.: Agile and adaptive governance in crisis response: lessons from the COVID-19 pandemic. Int. J. Inf. Manag. **55** (2020)
21. Karolak, D.W.: Global software development. Managing virtual teams and environments. IEEE Comput. Soc. (1998)
22. Kuhrmann, M., et al.: Hybrid software and system development in practice: waterfall, Scrum, and beyond. In: Proceedings of the 2017 International Conference on Software and System Process, pp. 30–39. ACM (2017)
23. Krishna, S., Sahay, S., Walsham, G.: Managing cross-cultural issues in global software outsourcing. Commun. ACM **47**, 62–66 (2004)
24. Liu, J.-W., et al.: The role of Sprint planning and feedback in game development projects: implications for game quality. J. Syst. Softw. **154**, 79–91 (2019)
25. Lindsjørn, Y., et al.: Teamwork quality and project success in software development. A survey of agile development teams. J. Syst. Softw. **122**, 274–286 (2016)
26. Mancl, D., Fraser, S.D.: COVID-19's influence on the future of agile. In: Paasivaara, M., Kruchten, P. (eds.) XP 2020. LNBIP, vol. 396, pp. 309–316. Springer, Cham (2020). https://doi.org/10.1007/978-3-030-58858-8_32
27. Misra, S.C., Kumar, V., Kumar, U.: Identifying some important success factors in adopting agile software development practices. J. Syst. Softw. **82**, 1869–1890 (2009)
28. Nguyen, T., Wolf, T., Damian, D.: Global software development and delay: does distance still matter? In: 3rd IEEE International Conference on Global Software Engineering Proceedings, pp. 45–54 (2008)
29. Ochodek, M., Kopczyńska, S.: Perceived importance of agile requirements engineering practices – a survey. J. Syst. Softw. **143**, 29–43 (2018)
30. Poth, A., Kottke, M., Riel, A.: The implementation of a digital service approach to fostering team autonomy, distant collaboration, and knowledge scaling in large enterprises. Hum. Syst. Manag. **39**, 573–588 (2020)
31. Purna Sudhakar, G.: A model of critical success factors for software projects. J. Enterp. Inf. Manag. **25**, 537–558 (2012)
32. Runeson, P., Höst, M.: Guidelines for conducting and reporting case study research in software engineering. Empir. Softw. Eng. **14**(2), 131–164 (2009)
33. Ram, P., et al.: Success factors for effective process metrics operationalization in agile software development: a multiple case study. In: IEEE/ACM International Conference on Software and System Processes, pp. 14–23 (2019)
34. Schön, E.-M., Thomaschewski, J., Escalona, M.J.: Agile requirements engineering: a systematic literature review. Comput. Stand. Interfaces **49**, 79–91 (2017)
35. Schwaber, K., Sutherland, J.: The Scrum Guide. https://www.scrumguides.org/scrum-guide.html
36. Senapathi, M., Drury-Grogan, M.L.: Refining a model for sustained usage of agile methodologies. J. Syst. Softw. **132**, 298–316 (2017)
37. Senapathi, M., Srinivasan, A.: Understanding post-adoptive agile usage. An exploratory cross-case analysis. J. Syst. Softw. **85**, 1255–1268 (2012)
38. Shah, Y., Raza, M., Ulhaq, S.: Communication issues in GSD. Int. J. Adv. Sci. Technol., 69–75 (2012)
39. Sinha, R., Shameem, M., Kumar, C.: SWOT: strength, weaknesses, opportunities, and threats for scaling agile methods in global software development. In: Proceedings of the 13th Innovations in Software Engineering Conference on Formerly Known as India Software Engineering Conference, pp. 1–10. ACM (2020)
40. Tam, C., et al.: The factors influencing the success of on-going agile software development projects. Int. J. Proj. Manag. **38**, 165–176 (2020)

41. Vanzin, M.-A., et al.: Global software processes definition in a distributed environment. In: 29th Annual IEEE/NASA Software Engineering Workshop, pp. 57–65. IEEE Computer Society (2005)

42. VersionOne; Collabnet. 14th Annual State of Agile Survey Report (2020). https://www.sta teofagile.com

43. Yin, R.K.: Applications of Case Study Research. Sage, Thousand Oaks (2008)

44. Yin, R.K.: Case Study Research. Design and Methods. Sage, Los Angeles (2009)

Agile Project Development Issues During COVID-19

Shariq Aziz Butt[1]([✉]) [iD], Sanjay Misra[2], Muhammad Waqas Anjum[3],
and Syed Areeb Hassan[4]

[1] University of Lahore, Lahore, Pakistan
Shariq2315@gmail.com
[2] Covenant University, Ota, Nigeria
sanjay.misra@covenantuniversity.edu.ng
[3] NCBA&E, Lahore, Pakistan
waqasch.065@gmail.com
[4] Superior University, Lahore, Pakistan
s.areebhassan@gmail.com

Abstract. Today's software development business is very much dynamic, and industries are changing their software development and requirements ways to meet the new changes in the environment. The environment also demands the rapid development and delivery of the product. In fast-changing environment and demands, agile methodology is the most useful development method. It gains fame due to its unique features that facilitate the software development more efficiently, client developer relation, rapid delivery of the product, and allow changes at any stage of the project, and client satisfaction. Currently, the world is suffering under a pandemic situation due to the COVID-19 disease. The disease spread with the close interaction with the human, for this reason many thousands of developers started working from home. The software industries working with the agile methodology are facing many issues to meet the development objectives. For this line of research, we have conducted this study on many different software industries using the agile methodology. In this study, we interviewed many developers using the questionnaire to determine the significant reasons for the failure of agile methodology during the current pandemic situation. We applied regression analysis, Cronbach alpha, and descriptive analysis statistical methods to the data set.

Keywords: Agile methodology · Software development · Work from home · Pandemic situation · Productivity in COVID-19

1 Introduction

The agile manifesto was introduced in 2001 by the research community working on software development to overcome the development issues [1]. Other development models have problems and no more beneficial for efficient software development. Therefore, to restore efficient software development, agile with its complete techniques were

© Springer Nature Switzerland AG 2021
A. Przybyłek et al. (Eds.): LASD 2021, LNBIP 408, pp. 59–70, 2021.
https://doi.org/10.1007/978-3-030-67084-9_4

introduced. Several studies [23–25] proved agile software development as more efficient amongst other software development methodologies. The agile presents various methodologies like Scrum, Extreme Programming, Crystal, and DSDM [2]. Amongst all available agile methods, Scrum [26] and extreme programming are more popular than others. Even Scrum is combined with other methods like Lean for getting better performance and productivity [27]. Scrum works in sprints, delivered within a specific time slot. In this present work, whenever we will talk about agile software development, although we consider all various agile methods, our main focus will be on the Scrum and Xtreme programming. The main objectives of Scrum methodology are as follows [3–5]:

- Divide the project into sprints, i.e., in user stories, to develop the modules.
- The most important goal of the agile model is the client's satisfaction. The main cause of the fame of agile methodology is the client preferences and satisfaction. It suggests client meeting on each sprint completion.
- Agile methodology prefers the change of request stage at any stage of the project/sprint by the client. It always welcomes the change request and gives priority to each request to complete in the net sprint.
- Agile enhances the collaboration of team/s to produce more productivity. It facilitates the team/s member/s to discuss the whole project and set a time slot and decide the project complexity.
- It improves the client and developer/s relation. It enables them to sit together and discuss work.
- It removes the obstacles of efficient software development.
- It delivers the project's modules rapidly.
- Urgently get feedback from the client on the developed sprint.
- Enable the self-organizing team/s.
- Less documentation.

During the COVID-19 pandemic, the whole world has been affected. Software development, which is a human-centric teamwork activity is also affected [28–30]. Agile software development, which is an effective way of producing software involves, customers in the development process [23, 24]. Scrum [26], one of the very popular agile methods, works in sprints and various types of meetings for solving issues and problems during development are part of that. During the COVID-19 pandemic, it became impossible to organize such physical meetings among software development team members and with customers. This present study aims to find issues while using agile software development during the pandemic situation. We identified the gaps and problems while evaluating the quality of the agile development methodology during this pandemic period. We don't found any study in the literature that highlights the difficulties in agile software development during COVID-19 pandemic.

The main objective of the study is to find the major factors that influence the agile productivity in the pandemic situation. In this line of research, we design some hypotheses as research questions and conduct a survey. We adopted three statistical techniques on the collected data set and concluded the study with the result's findings. We reveal that in the pandemic situation, the agile reduced the software's productivity.

The structure of the paper is as follows. A survey modelling adopted for this study is provided in Sect. 2. Section 3 contains the experimental design and methodology, and Sect. 4 presents the statistical results applied to the collected data set and reveals the findings of the study.

2 Survey Modelling

The main focus of the study is to explore the use of agile methodology during the pandemic situation and its impact on productivity either in positive or negative ways. It means the pandemic situation boosted the productivity of development or reduced it. If boosted then how and if reduce, then what are the causes of it. In this line of research, we conducted this survey, we designed the 10 Hypotheses mentioned in Table 2 [10], and the research model is shown in Fig. 1. Each hypotheses is associated with a variable that predicts the impact of the pandemic situation on these variables. The main variables of the study are (1) Coordination, (2) Time, (3) Client Meeting, (4) Cost, (5) Remote Working, and (6) Work Satisfaction. All hypotheses set for the study are evaluated and measured under these defined variables. The variable coordination is covering the Hypotheses H3 and H9, as these Hypotheses are directly influenced by the coordination of team/s and software developer/s [3, 6, 7]. Variable cost impact on Hypotheses H4 and H5 cost increase relation is to the late delivery of sprints and work from home that makes the developers lazy. The variable Client Meeting has a strong relationship with the Hypotheses H1, H2, and H10 [8–10]. The reason is the client has a concern with the product's outcome, frequent meetings with clients such as agile always support, and agile effectiveness due to COVID-19. The variable time has a concern with the Hypotheses H4, H5, and H7. Furthermore, the variables Remote working and Work satisfaction have an impact on Hypotheses H6, H8, and H10 [8, 11, 12]. We conducted a survey amongst different software industries and many professionals working for many years in the agile model. The data gathering and sourcing for this analysis are mentioned in Table 1.

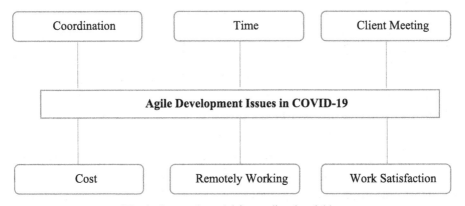

Fig. 1. Research model for predicted variables

Table 1. Specification of data, survey goals, data sourcing, and gathering.

Purpose of the study	To recognize the main issues while using agile development during the COVID-19
Subject area	Agile effectiveness during a pandemic
Main research Question	To find the main problems and research questions in agile project development during the pandemic period and how to solve it?
More specification subject area	Pandemic for agile in software industries
Data type	Hypotheses
How data is acquired	From analysis of hypotheses
Data format	Examined and statistical data
Investigational factors	Hypotheses model is created with the help of software developers who are working in the pandemic situation at home
Experimental features	Agile work at home effectiveness
Data source location	Software Industries

Table 2. Hypotheses

H 1	Do you think quality agile methodology still has a positive outcome in the pandemic situation?
H 2	Do you think the COVID-19 has a negative effect on software development in agile?
H 3	Do you think that the Pandemic situation creates issues for team/s coordination?
H 4	Do you agree that the cost of projects increases during the work home?
H 5	Do you agree that the projects/Sprints deliver late to the client while working from home?
H 6	Do you think that productivity decrease during the COVID-19?
H 7	Do you think an influence of work pressure in COVID-19 has a positive effect?
H 8	Do you feel that the work satisfaction rate among the developer/s becomes low?
H 9	Do you agree with that the team/s face issues for the remote meeting in the pandemic situation?
H 10	Do you feel that the meeting with the client is suffering during work from home?

3 Experimental Design and Methodology

For the survey conduct, we visited 10 software industries to get responded for our hypotheses as shown in the Appendix. Here we use the inclusion and exclusion criteria for software industry. We only include the software industries which are using the agile model for software development in the current pandemic situation, and the others are excluded from our survey results that are not using agile for development. Therefore, to conduct the survey, we selected four software industries using agile for software development during the pandemic situation (COVID-19). Table 3 is presenting the companies selected for the survey session. We selected the companies that are using the agile model for development. Only those companies are selected which have international offices, a high number of employees, and developing some software applications [13, 14]. We include only software industries that have some high years of experience using agile, therefore, they can respond more accurately to our hypotheses. The survey modeling is summarized in Table 4.

Table 3. Survey industries

Companies	No. employees	Type of services	Location	Sub-locations
Company 1	300–400	IT business solutions	Pakistan	Dubai
Company 2	250–300	Software applications development	Pakistan	UAE
Company 3	200–250	Govt. software applications	Pakistan	Australia
Company 4	200–220	IT Solutions consultancy	Pakistan	UK

Table 4. Survey modeling

Objective	Use of agile during the pandemic situation in software industries
Supporting audience	Software Developers, Team Leaders, and Project Managers
Data collection mode	Questionnaire
No of Industries for survey	4
Total questions	10
Participants	250+

4   Results and Analysis

We arranged this survey session in many reputed software industries. A 250+ number of participants joined the survey sessions and responded to the questions. We send all EMPLOYEES to the software industries, but only the mentioned participants respond the hypotheses. For the respond number of days given to participants. The Hypotheses are arranged according to their suggestions. We eliminate the responses that are mistakable and not appropriately filled. We also did't consider the missing data in our results as output. Instead of starting and applying any statistical technique on data, we try to check the validation of the Hypotheses through the Cronbach Alpha test. We apply this technique to the variables mentioned in the research model. The analysis of this test on Composite Reliability and Cronbach Alpha are ≥ 0.7. Additionally, the Average Variance

Extract is also more than 0.60. This result shows that the variables with the mentioned Hypotheses are statistically reliable and legitimate for this study [20, 21]. The outcome is shown in Table 5.

Table 5. Cronbach Alpha.

Predictive variable	Cronbach's Alpha	Composite reliability	Average variance extract
Coordination	.62	.84	.72
Time	.92	.82	.65
Cost	.63	.74	.73
Work satisfaction	.61	.81	.63
Client meeting	.83	.73	.64
Remotely working	.71	.90	.70

After the Cronbach Alpha results, we used two statistical techniques on the data set generated through the survey results. The descriptive analysis technique applied to Hypotheses [22] is shown in Table 6. The means and mode of all Hypotheses are 1 and the calculations of the SES and SEK are 0.824 and 0.152, respectively. Table 5 presents that almost 81% of developers stated that the agile work quality for efficient software development has a negative impact rather than a positive under the response of Hypotheses H1 and H2. The participants stated that agile has a significant role in software development in positive ways, but now, due to the pandemic situation, the agile values go down. In the agile, coordination of team/s plays a significant role because the coordination supports the participants to work together for the project. It also makes the team/s dependent on each other.

In our study, Hypotheses H3 reveals that the coordination of team/s and participants within the team becomes low. 80% of developers stated that the coordination is almost non-existent due to the pandemic situation. Everyone is at home; thus, the discussions about the user stories, and complexity of the project are about to zero. Just some of the participants take part in the discussion and decide parameters about the project. During the agile project development in the pandemic situation, the project's cost under the evaluation of Hypotheses H4 stated that the cost increase. The main causes of the increase of cost are: not properly responded by the developers during the sprints, low working hours of developers, there was no work pressure on them, there is no sprint Scrum meeting and mentally disturbed due to pandemic situation. On the other hand, agile has worked for cost management during project development. It improved cost-saving of project development. A few studies reported that agile has no control over cost, and its features increase the cost of projects. In our study, 75% of participants revealed that the cost becomes increased due to COVID-19. As we have done surveys individually with all participants, therefore nobody knows about the response of others. This methodology reveals the outcomes against the Hypotheses H5, 84% used for the evaluation of variables Remotely Working, it greatly impacts on the agile project development. The participant does not know about the responses of each other, so everyone stated that other developers worked slowly from home. On asking the developer/s, they gave some non-professional reasons for the late delivery of sprints and not attend the online meeting. Their remote working behavior changes the working practice of agile for any project that makes the agile less effective. Anyhow, the literature shows that the agile builds professional attitudes and behaviors among the developer/s. The Hypotheses H6 and H7 are evaluated using the variable time; the survey results reveal that the developer/s take more time to complete the user stories that are not enough complex. Due to sprint deliveries being late at the client's, ultimately the project time increases. While in the pandemic situation COVID-19, every stakeholder is shifting to the E-business but the agile delivers projects late. Participants 75–65% stated that agile is not proficient in the work from home scenario. It even becomes the worst for development rather than the traditional development approaches. Hypotheses H8 and H9 are designed and used to examine how the developer/s feels about the work at home as per their satisfaction level. As agile increases the work satisfaction level of developer/s due to its unique features that support to increase productivity. Due to stress and emotional, mental instability affects their productivity level. Work pressure and home life have a strong conflict which is the leading cause of less productive team/s. Agile always supports client satisfaction as the priority of the software industry, but in the current pandemic situation human can't meet physically, therefore the meetings with the client suffers a lot. The developer/s stated that the meetings are less scheduled due to time issues with the client, low internet speed, and availability of both to discuss the project. Most of the feedback from the client was taken via email or on story cards. Hypotheses H10 revealed 77% of participants stated that the client suffered a lot in pandemic situations [15–17].

Table 6. Descriptive analysis of hypotheses.

Hypotheses	Mean	SE	SD	SK	KU
H 1	0.81	0.52	0.56	−0.7	1.5
H 2	0.78	0.46	0.32	−0.8	−1.5
H 3	0.82	0.46	0.54	−0.7	−2.0
H 4	0.76	0.51	0.30	−1.1	−.13
H 5	0.84	0.32	0.37	−.30	−0.7
H 6	0.76	0.41	0.46	−.31	−2.7
H 7	0.64	0.30	0.47	−.06	−1.2
H 8	0.77	0.41	0.58	−.21	−1.9
H 9	0.73	0.31	0.33	−1.5	3.5
H 10	0.77	0.42	0.43	−0.6	2.8

We have done the Regression Analysis test on the data set to find the Anova, coefficients and Model Summary. We used regression analysis for the quality outcomes as shown in Table 7. In our study, the regression analysis test is used to explain the connection between the defined variables and the impact of Hypotheses on the agile use during

Table 7. Regression analysis.

	Model summary		ANOVA		Un-standardized coefficients		Standardized coefficients	t	Sig. t
	Multiple R	R^2	F	Sig.F	B	SE	ß		
H 1	0.82	.56	77.62	0.00	.86	.09	0.80	9.99	0.00
H 2	0.73	.37	55.56	0.00	.88	.06	0.73	6.35	0.00
H 3	0.80	.58	55.78	0.00	.70	.09	0.80	6.90	0.00
H 4	0.77	.16	27.10	0.00	.45	.18	0.77	4.37	0.00
H 5	0.80	.29	13.32	0.00	.66	.07	0.80	5.93	0.00
H 6	0.79	.45	41.40	0.00	.60	.06	0.79	5.53	0.00
H 7	0.66	.43	40.24	0.00	.56	.10	0.66	4.59	0.00
H 8	0.75	.36	28.37	0.00	.63	.10	0.75	4.02	0.00
H 9	0.78	.40	43.42	0.00	.78	.27	0.78	5.58	0.00
H 10	0.76	.52	55.74	0.00	.69	.06	0.76	6.70	0.00

the pandemic situation. We adopted model summary in the regression test which the Multiple R is the square of the R square, and it picks an association coefficient amongst the picked predicted variables. The Multiple R has the highest value of the Hypotheses H1, 80%. As well, the lowest value is Hypotheses H7 with 67%. The values show that the agile during the COVID-19 has a negative impact on software development. It reduces its fame in the software industry. In the regression analysis, the correlation coefficient is continuously going between +1 and −1. Correlation coefficient with respect to (0 to .3) is considered as a weak coordinate association, between (.3 to .7) that reveals the workable relationship between (0.7 to 1) exhibits a solid association [17–19, 29]. In the test, R Square (R2) is stated as the weight to predict the variables defined in the survey model and measure the effectiveness of Hypotheses under the model. However, the Anova F-test in the regression analysis shows that the impact of predicted variables by the defined Hypotheses against it. There is also a need that the F-Sig test value should be higher than the 0.05, thus in our case, the F-Sig value is (p-esteem <0.05). The B coefficient shows the H4 to 0.45. The ß estimation of each predicted variable is the same as Multiple R regard. It is utilized to identify the invalid or null theory. The ß is coupled with t-value and the significance of t-regard to confirm whether the coefficient is below and more than 0 and subject to this, ß can be either + or −. The prediction variables were established based on the un-standardized beta coefficient for the Hypotheses.

5 Discussion and Analysis

The results reveal that the software industries using the agile methodology during the pandemic situation have a negative impact with 82% reducing the influence of agile model. The agile does not work effectively during the COVID-19 pandemic. Hypotheses H7 has the lowest value with 66% that shows there was no work pressure, so it impacts the productivity of the project development using the agile methodology, decreased in the COVID-19 period. There is a massive downfall in software productivity compared to normal. As the agile has the fame to control the cost and time during the project development, but in the case of COVID-19 the cost and time of projects during the situation become increased by 77 and 80%. The other cause is less coordination among the team/s and developer/s during the work at home. As the COVID-19 places everybody at home, therefore the individual have less ambitious and work satisfaction. 75% of participants stated that they feel less satisfied with their input on any user story. The major reason is spending time with family, no official environment to work, the stress of CORONA virus, health and mental stress, no work pressure, and less talk with others. During the COVID-19 period, the most suffered individual/s clients 76%, who faced a delay in the project delivery time and cost increase, the output of user story is not according to need, fewer meetings with the developer/s, and less satisfaction of work. Figure 2 is explaining the overall results of the Hypotheses.

The threats to the validity of the study are as follows:

Internal Threats
This threat refers to whether or not something influenced the results of the study.

Selection: The selection of participants is a threat due to the pandemic situation. We were unable to contact more accurate persons for our survey study. Therefore, we set criteria for the selection of parson who has a background of development with agile and can be a developer, project manager, team lead, and principal software design and architect.

Techniques: To remove any error from the results, we only selected the statistical methods that are discussed in the literature for survey studies.

External Threats
The survey study has been validated by the participants. As we don't include any missing or non-satisfactory response of any hypotheses in our data set.

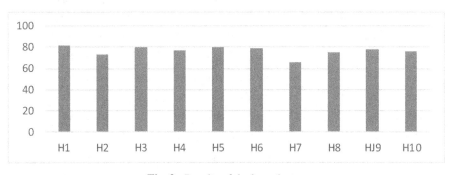

Fig. 2. Results of the hypotheses.

6 Conclusion

The agile methodology is the most famous development model for efficient software development. Its unique techniques make it more viable for adoption in the software industry. However, during the pandemic situation, the agile methodology faced so many issues that impact software development negatively. We used three statistical techniques on the collected data set to reveal the study results. The most factors in agile not useful for the pandemic situation are work from home, less satisfaction of work, fewer meetings with the client, mental and health stress, less work pressure, and an increase in cost and time. Besides, this examination is remarkable in a few different ways: (1) the survey utilized recently approved scales, which we re-approved using both head part investigation and corroborative factor examination; (2) the information was dissected utilizing profoundly advanced strategies (for example, basic condition demonstrating), which seldom have been used in programming building research; (3) the examination explores a rising wonder, giving convenient guidance to associations and experts. We selected four software industries and 250+ participants, but it can be increased with more industries and participants to participate in this study for more data set and accurate results. This limited our study to engage more and more people in our survey. We trust that this examination reuses more exploration of how programming improvement is influenced by emergencies, pandemics, lockdowns, and other antagonistic conditions.

The limitation of the study is that due to the pandemic situation, we are unable to engage more participants from software companies in survey data collection. However, this survey study reveals that agile needs some amendments to make it more proficient for development. These amendments make it more adaptable and feasible in the case of a pandemic.

Appendix

List of Companies.

Companies	No. employees	Type of services	Location	Sub-locations
Company 1	300–400	IT business solutions	Pakistan	Dubai
Company 2	250–300	Software applications development	Pakistan	UAE
Company 3	200–250	Govt. software applications	Pakistan	Australia
Company 4	200–220	IT solutions consultancy	Pakistan	UK
Company 5	150–200	Software applications development	Pakistan	Romania
Company 6	200–300	Learning applications development	Pakistan	Austria
Company 7	200–350	Health applications development	Pakistan	UK
Company 8	150–230	IT solutions consultancy	Pakistan	UAE
Company 9	200–250	Govt. software applications	Pakistan	USA
Company 10	150–300	Business solutions	Pakistan	UAE

References

1. Steghöfer, J.P., Knauss, E., Alégroth, E., Hammouda, I., Burden, H., Ericsson, M.: Teaching agile-addressing the conflict between project delivery and application of agile methods. In: 2016 IEEE/ACM 38th International Conference on Software Engineering Companion (ICSE-C), pp. 303–312. IEEE (2016)
2. Meyer, B.: Making sense of agile methods. IEEE Softw. **35**(2), 91–94 (2018)
3. Khalid, A., Butt, S.A., Jamal, T., Gochhait, S.: Agile scrum issues at large-scale distributed projects: scrum project development at large. Int. J. Softw. Innov. (IJSI) **8**(2), 85–94 (2020)
4. Martin, A., Anslow, C., Johnson, D.: Teaching agile methods to software engineering professionals: 10 years, 1000 release plans. In: Baumeister, H., Lichter, H., Riebisch, M. (eds.) XP 2017. LNBIP, vol. 283, pp. 151–166. Springer, Cham (2017). https://doi.org/10.1007/978-3-319-57633-6_10
5. Butt, S.A.: Study of agile methodology with the cloud. Pac. Sci. Rev. B Humanit. Soc. Sci. **2**(1), 22–28 (2016)
6. Fuchs, C.: Adapting (to) agile methods: exploring the interplay of agile methods and organizational features (2019)
7. Wińska, E., Dąbrowski, W.: Software development artifacts in large agile organizations: a comparison of scaling agile methods. In: Poniszewska-Marańda, A., Kryvinska, N., Jarząbek, S., Madeyski, L. (eds.) Data-Centric Business and Applications. LNDECT, vol. 40, pp. 101–116. Springer, Cham (2020). https://doi.org/10.1007/978-3-030-34706-2_6
8. Tessem, B.: The customer effect in agile system development projects. A process tracing case study. Procedia Comput. Sci. **121**, 244–251 (2017)
9. Butt, S.A., Abbas, S.A., Ahsan, M.: Software development life cycle & software quality measuring types. Asian J. Math. Comput. Res. **11**, 112–122 (2016)
10. Butt, S.A., Jamal, T.: Frequent change request from user to handle cost on project in agile model. Proc. Asia Pac. J. Multi. Res. **5**(2), 26–42 (2017)
11. Kim, S.I., Lee, J.Y.: Walk-through screening center for COVID-19: an accessible and efficient screening system in a pandemic situation. J. Korean Med. Sci. **35**(15), e154 (2020)
12. Tariq, M.I., Diaz-Martinez, J., Butt, S.A., Adeel, M., De-la-Hoz-Franco, E., Dicu, A.M.: A learners experience with the games education in software engineering. In: Balas, V.E., Jain, Lakhmi C., Balas, M.M., Shahbazova, Shahnaz N. (eds.) SOFA 2018. AISC, vol. 1222, pp. 379–395. Springer, Cham (2021). https://doi.org/10.1007/978-3-030-52190-5_27
13. Janssen, M., van der Voort, H.: Agile and adaptive governance in crisis response: lessons from the COVID-19 pandemic. Int. J. Inf. Manage. **55**, 102180 (2020)
14. Asare, A.O., Addo, P.C., Sarpong, E.O., Kotei, D.: COVID-19: optimizing business performance through agile business intelligence and data analytics. Open J. Bus. Manage. **8**(5), 2071–2080 (2020)
15. da Camara, R., Marinho, M., Sampaio, S., Cadete, S.: How do agile software startups deal with uncertainties by COVID-19 pandemic? arXiv preprint arXiv:2006.13715 (2020)
16. Goel, S., et al.: Resilient and agile engineering solutions to address societal challenges such as coronavirus pandemic. Mater. Today Chem. **17**, 100300 (2020)
17. Ralph, P., et al.: Pandemic programming: how COVID-19 affects software developers and how their organizations can help (2020). arXiv preprint arXiv:2005.01127
18. Ratner, B.: The Correlation Coefficient: Definition, DM Stat-1 Articles, vol. 11a (2007)
19. Lawal, B.: Applied Statistical Methods in Agriculture, Health, and Life Sciences. Springer, Cham (2014). https://doi.org/10.1007/978-3-319-05555-8
20. Alzoubi, H., Yanamandra, R.: Investigating the mediating role of information sharing strategy on agile supply chain. Uncertain Supply Chain Manage. **8**(2), 273–284 (2020)

21. Iqbal, T., Jajja, M.S.S., Bhutta, M.K., Qureshi, S.N.: Lean and agile manufacturing: complementary or competing capabilities? J. Manuf. Technol. Manage. **31**(4), 749–774 (2020)
22. Kumar, R., Singh, K., Jain, S.K.: Agile manufacturing: a literature review and Pareto analysis. Int. J. Qual. Reliab. Manage. **37**, 207–222 (2019)
23. Patel, A., Seyfi, A., Taghavi, M., Wills, C., Na, L., Latih, R., Misra, S.: A comparative study of agile, component-based, aspect-oriented and mashup software development methods. Tehnicki Vjesnik **19**(1), 175–189 (2012)
24. de la Barra, C.L., Crawford, B., Soto, R., Misra, S., Monfroy, E.: Agile software development: it is about knowledge management and creativity. In: Murgante, B., et al. (eds.) ICCSA 2013. LNCS, vol. 7973, pp. 98–113. Springer, Heidelberg (2013). https://doi.org/10.1007/978-3-642-39646-5_8
25. Pham, Q.T., Nguyen, A.V., Misra, S.: Apply agile method for improving the efficiency of software development project at VNG company. In: Murgante, B., et al. (eds.) ICCSA 2013. LNCS, vol. 7972, pp. 427–442. Springer, Heidelberg (2013). https://doi.org/10.1007/978-3-642-39643-4_31
26. Mundra, A., Misra, S., Dhawale, C.A.: Practical scrum-scrum team: way to produce successful and quality software. In: 2013 13th International Conference on Computational Science and Its Applications, pp. 119–123. IEEE (2013)
27. Correia, A., Gonçalves, A., Misra, S.: Integrating the scrum framework and lean six sigma. In: Misra, S., et al. (eds.) ICCSA 2019. LNCS, vol. 11623, pp. 136–149. Springer, Cham (2019). https://doi.org/10.1007/978-3-030-24308-1_12
28. Mishra, A., Misra, S.: People management in the software industry: the key to success. ACM SIGSOFT Softw. Eng. Notes **35**(6), 1–4 (2010)
29. Fernández-Sanz, L., Gómez-Pérez, J., Diez-Folledo, T.I., Misra, S.: Researching human and organizational factors impact for decisions on software quality. In: Proceedings of the 11th International Conference on Software Engineering and Applications, pp. 283–289 (2016)
30. Fernández-Sanz, L., Misra, S.: Influence of human factors in software quality and productivity. In: Murgante, B., Gervasi, O., Iglesias, A., Taniar, D., Apduhan, Bernady O. (eds.) ICCSA 2011. LNCS, vol. 6786, pp. 257–269. Springer, Heidelberg (2011). https://doi.org/10.1007/978-3-642-21934-4_22

Achieving Agility in IT Project Portfolios – A Systematic Literature Review

Joseph Puthenpurackal Chakko[1](✉) (iD), Tim Huygh[2] (iD), and Steven De Haes[1] (iD)

[1] Antwerp Management School, Antwerp, Belgium
josephpc@yahoo.com, steven.dehaes@uantwerpen.be
[2] Open University of the Netherlands, Heerlen, The Netherlands
tim.huygh@ou.nl

Abstract. Over the past two decades, enterprise IT functions have enjoyed continued success in projects using agile development methods. However, the lack of ample empirical research on achieving portfolio level agility can potentially inhibit their ability to effectively govern IT investments while scaling agile practices to derive more significant benefits. This study examines the impact of agile delivery efforts on project portfolio management at the enterprise level and identifies approaches adopted to foster agility in portfolio practices. We conducted a systematic literature review to explore existing scientific knowledge around agile methods and portfolio management in an enterprise IT context. An analysis of the 21 primary studies found relevant to this research identified six portfolio management aspects impacted by agile delivery practices and a variety of approaches adopted to support them. While these identified portfolio management aspects guide practitioners on areas to focus on while scaling agile efforts across an enterprise, the specific practices/approaches observed present opportunities to consider within their respective organizational contexts. Portfolio processes need an exploratory focus to sense environmental change to support agility, utilize a systems-thinking approach for a holistic view of potential interactions within and across portfolio components, and consider the effect of existing organizational processes to support portfolio agility. This study contributes to academic knowledge by synthesizing current knowledge on how portfolio management contributes to IT agility while incorporating agile delivery efforts and by identifying a set of future research directions in this space.

Keywords: Agile methods · IT agility · Portfolio management · Systematic literature review

1 Introduction

Today's enterprises face increasing pressure from the complex dynamics of their markets, forcing them to critically examine their business models to stay ahead of their competition [1]. As a result, information technology (IT) capabilities are being called on to enable options to drive business model innovation [2]. IT agility, the two-dimensional capability to sense and respond to changing IT environments, enables IT functions

© Springer Nature Switzerland AG 2021
A. Przybyłek et al. (Eds.): LASD 2021, LNBIP 408, pp. 71–90, 2021.
https://doi.org/10.1007/978-3-030-67084-9_5

to influence their *"position to impact strategic business decisions"* [3]. Enterprise IT functions are increasingly adopting agile delivery practices leading to improved project delivery efficiency, product quality, stakeholder satisfaction, and project performance [4]. The success of agile methods has led enterprises to consider applying them at a larger scale, with support from several scaling frameworks introduced in the practitioner space to align these development efforts to business strategies.

There is extensive research on the use of agile methods to deliver project outcomes. In comparison, the investigation into scaling agile practices to the enterprise level has been less prevalent. Given that agile practices focus on shorter planning/delivery cycles that continually adapt to and align with evolving customer needs, portfolio management structures and processes need adjustments to preserve IT agility. Although portfolio management is considered a critical aspect of large-scale agile development [5], there is limited research conducted into the structures and processes supporting portfolio activities governing agile environments [6].

This literature review brings together our current understanding of how project portfolios enable IT agility while incorporating agile delivery efforts and is a unique contribution by being the first to explore the current state of knowledge at the intersection of portfolio management and agile practices. This review aims to understand better how enterprises can achieve agility in their IT portfolio management process. It identifies impacts on portfolio areas from agile delivery methods and describes approaches adopted to address these impacts. This review is part of a broader effort to design an agile portfolio management framework, and the review findings will form the basis for future studies on portfolio management practices enabling IT agility.

Section 2 of this paper provides a conceptual background to facilitate the literature review, and Sect. 3 describes the systematic literature review process. The review results in Sect. 4 present a set of themes crucial to achieving agility in IT portfolios. Section 5 discusses the implications of these findings, while Sect. 6 concludes by summarizing the contributions and calling out future research directions.

2   Background

This section explores key concepts in IT project portfolio management and agile software development principles to frame the literature review.

2.1   Project Portfolio Management

Levine [7] describes project portfolio management (PPM) as a set of practices binding traditional operations management and project management disciplines to ensure project contributions are maximized and aligned to enterprise success. PPM is the means to realize enterprise strategies [8] by screening, selecting, continuously prioritizing, and allocating resources to projects in line with strategic priorities [9]. Extensive research conducted on portfolio practices across various enterprises [10] identified maximizing portfolio value, achieving a balance in the mix of projects, and ensuring alignment to business strategies as portfolio management's goals. Müller et al. [11] categorize PPM activities into three groups of top-down methods that (1) align projects with business strategy and prioritizes them (portfolio selection), (2) continuously monitor and communicate project priorities and progress at the portfolio-level (portfolio reporting) and (3) make rational and objective choices to accelerate, kill or reprioritize projects within the portfolio (portfolio decision-making).

McFarlan [12] suggested using the portfolio model to manage overall risk exposures for information technology (IT) projects in a manner analogous to applying the modern portfolio theory [13] to an investment portfolio of diversified financial securities. The US General Accounting Office [14] recommends a portfolio investment approach to select, control, and evaluate IT projects by defining and applying a set of decision criteria across benefits, costs, and risks associated with the competing project investment options. Maizlish and Handler [15] describe IT PPM as "a combination of people, processes, and corresponding information and technology that sensed and responded to change by reprioritizing/rebalancing investments and assets, value-based risk assessment of existing assets, eliminating redundancies while maximizing reuse, optimal resource allocation, and continuous monitoring & measuring."

Project portfolio management is an essential IT governance practice [16] to realize the expected business value from IT-enabled investments by aligning business objectives and IT strategies. A review of project governance literature [17] identifies two perspectives for project governance – an external system of governance that focuses on centralized monitoring and controls to ensure project outcomes stay aligned to strategic objectives and an internal one that builds organizational capabilities to achieve shared project goals. Kujala et al. [18] propose a framework to support project governance across six dimensions (goal setting, incentives, monitoring, coordination, decision-making, and capability building).

2.2 IT Agility and Agile Software Development

Leonhardt et al. [3] view IT agility as a capability that, on the one hand, proactively senses and assesses emergent developments and opportunities, and on the other hand, maintains an IT landscape that enables swift response and adaptation to the changing business needs. The 14th Annual State of Agile Survey indicates that enterprises adopt agile software development methods to accelerate their software delivery (71%), to enhance their ability to manage changing priorities (63%), and to increase productivity (51%) [19]. Agile methods, like Scrum [20] based on agile values/principles [21] and concepts of empirical process control, were conceptualized to improve the way software development projects are organized and executed.. They use iterative and incremental delivery of project results with self-organized cross-functional teams using patterns of actions like daily stand-up meetings for team coordination and frequent reviews with close customer contact [22]. Studies indicate that agile delivery methods positively impact efficiency, stakeholder satisfaction, and perception of overall project performance [4] and reduce overall project delivery timelines [23].

2.3 Agile Portfolio Management

Agile portfolio management extends existing portfolio activities [11] by connecting agile software development teams to enterprise strategies allowing for rapid reconfigurations of portfolio components in response to changing environments, thereby enabling IT agility [3]. This view of applying agile principles to the portfolio level is consistent with Krebs' approach to dynamically manage portfolios using flexible financial models [24] or the Scaled Agile Framework's direction for lean portfolios [25]. Traditional portfolio management often takes a linear, top-down approach and focuses on long-term-planning and control [26], while agile principles highlight the need to be iterative and responsive to change [27]. There is a difference in the granularity of planning (informal vs. formal and *a priori* vs. evolutionary) within agile efforts [28], particularly while considering resource allocations, ranging from smaller projects to complex enterprise-level portfolios [29]. Elements like team autonomy and diversity advocated in agile methods indicate a people-centric approach [30] compared to the more resource-oriented view of traditional portfolio management.

An exploration of existing agile scaling frameworks, like the Scaled Agile Framework[1] (SAFe), Large Scale Scrum[2] (LeSS), and Disciplined Agile Delivery[3] (DAD), indicated little consistency in their recommendations to scale agile efforts to the portfolio level [6, 31, 32].

Table 1 lists literature reviews identified during the preliminary search relating to large-scale agile practices. Although none of these reviews directly address portfolio-level impacts while scaling agile practices, they provided relevant background information to identify themes for use during the data synthesis process.

Table 1. Past literature reviews identified

No.	Reference	Focus area
1	Lappi et al. [33]	Mapping traditional governance practices to agile contexts using the project governance model proposed by Kujala et al. [18]
2	Alqudah and Razali [32]	Comparing the roles and practices of six common frameworks to scaling agile practices
3	Dikert et al. [34]	Identifying challenges/success factors for large-scale agile transformations
4	Ahmad et al. [35]	Explores the use of Kanban in support of software engineering practices

3 Review Method

The aim of a literature review is "to map and to assess the existing intellectual territory", to incorporate a degree of rigor into the inquiry process, and to develop a comprehensive

---

[1] See https://scaledagileframework.com for more details.

[2] See https://less.works for more details.

[3] See https://pmi.org/disciplined-agile for more details.

knowledge-base for practitioners from a range of past studies [36]. Established practices of conducting literature reviews in the information systems space [37, 38] guide the protocols described in this section. The literature review protocol covers detailed research questions, identifying literature sources, search strategy, inclusion, exclusion, and quality assessment criteria, processes to extract and synthesize data from identified studies, and reporting findings.

The research objective to better understand how enterprises can achieve agility in their IT portfolio management process is addressed through two research questions that guide and direct the review.

RQ1 – How have agile software delivery methods impacted existing enterprise portfolio management practices
RQ2 – What approaches/practices have enterprises adopted to achieve agility in meeting portfolio objectives?

3.1 Inclusion and Exclusion Criteria

Based on the various aspects of inquiry derived from the research questions, Table 2 provides guidance for inclusion and exclusion decisions for this review to ensure that only studies relevant to the research questions are selected.

Table 2. Inclusion and exclusion criteria

RQ Aspect	Inclusion examples	Exclusion examples
Portfolio Management	IT Portfolio Management; Governance of IT project investments; Multi-project practices	Financial securities, product/service portfolios; Focus on IT aspects like strategy & planning, architecture, process & performance, capabilities, culture
Agile	Lean/agile software development methods used in teams and product groups; Use of agile scaling frameworks	Agile manufacturing, contracting or supply chain practices; Descriptions of or experiences with specific agile methods
Enterprise IT Context	Multiple s/w dev teams, Solution delivery against business plans; Structures, and processes for project delivery	Individual or single team settings; Non-IT related business processes (e.g., training methods, business processes)
Empirical	Qualitative & quantitative studies; Peer-reviewed journal articles & conference papers	Conceptual papers, grey literature, vendor/analyst whitepapers, and other non-academic sources

3.2 Data Sources and Search Strategy

This study's topic cuts across research in information systems, computer science, software engineering, and project management. The search used six electronic databases – ACM Digital Library, AIS Electronic Library, IEEE Xplore, Elsevier ScienceDirect, Scopus, and Web of Science – to accommodate topics' breadth.

The preliminary search process used multiple combinations of terms for inclusion (like *"agile," "agility," "lean," "large scale," "enterprise," "governance," "scaling," "transformation," "portfolio," "project,"* and *"software"*) and exclusion (like *"manufacturing," "supply chain,"* and *"contract"*) to observe patterns and relevance in search outputs and to evolve suitable search criteria for the review. The final search string applied against the Title, Abstract, and Keywords in each database[4] is as follows:

> *"agile" AND "portfolio" AND ("software" OR "information" OR "governance" OR "scale" OR "lean")*

3.3 Study Selection Process

After removing duplicate results from the initial search (Stage 1), citations (n = 516) are loaded into an EndNote library. The metadata to support retrieval and inclusion decisions is maintained and tracked as review records in Microsoft Excel worksheets. The study selection process spans three stages, as shown in Fig. 1.

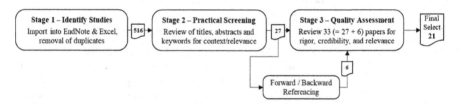

Fig. 1. The multi-stage study selection process

During Stage 2, the reviewer examined titles, abstracts, and keywords of each selected paper using the inclusion and exclusion criteria (described in Table 2) to establish their relevance to this review. After removing obvious exclusions (based on publication channels, research topics, and non-empirical papers), abstracts were scanned for factors such as domain under investigation (describing portfolio management in the contexts other than IT, like financial securities or product/service portfolios) and IS focus area (related to IT strategy & planning, IT architecture, IT processes, IT capabilities, culture, and performance instead of IT portfolio management or multi-project practices) to identify papers that need to be eliminated from the review process. There were 27 papers selected at the end of this second stage.

---

[4] The search string was implemented in the syntax unique to each database. Database searches were conducted in early June 2020.

The review employs a forward and backward snowballing process [39] using Google Scholar[5] to examine the citations and the references included in the 27 selected papers. This exercise identified six additional studies relevant to the topic that did not appear in the search process.

Stage 3 performs a detailed full-text review of the 33 selected papers (27 papers included from Stage 2 and six papers from the snowballing process) for their methodological rigor, the credibility of their results, and the relevance of their findings based on quality assessment criteria guided by recommendations from Kitchenham and Charters [37]. Of the six criteria identified, the first one ('*The research objective of this study is pertinent to the review*') is used to eliminate studies where the objectives do not map to the review's objectives. The other five criteria describe factors relating to rigor, credibility, and relevance of the studies and are scored on a 5-point Likert response format from 'Strongly Disagree' (1) to 'Strongly Agree' (5) based on each study's overall quality and strength of evidence. Studies with mean scores lower than 2.5 (indicating quality issues across most criteria) were removed from further review.

Based on the quality assessment results shown in Table 3, five studies were found irrelevant to the review. Seven studies of insufficient quality were eliminated, resulting in a final selection of 21 papers for further review.

3.4 Data Extraction and Synthesis of Findings

The final 21 studies selected for this review forms the input to the data extraction and thematic synthesis [40] stage (see Appendix for the list of selected studies). The initial data extraction captured bibliographic (author, year, source, and type of publication) and contextual (the focus area, research objective, research design, study setting, data collection & analysis methods, findings, and conclusions) information for each paper into a structured Microsoft Excel spreadsheet. The EndNote library, including the associated full-text files, was imported into nVivo to perform content analysis.

Each study is coded in nVivo for its setting, theoretical basis, findings, and results using a set of code families based on the two research questions ('impact to portfolio practices' and 'approaches to support agile practices') and on concepts from the theoretical framework like portfolio management [10, 11], agile principles [21] and project governance [33]. These codes were reviewed and organized to represent conceptual hierarchies that translated into themes.

3.5 Threats to Validity

We use factors such as internal validity, construct validity, external validity, and conclusion validity [41] to explore threats to this review's validity. Since this review aims to identify those portfolio management aspects impacted by agile delivery methods and not to determine any causal factors, threats relating to *internal validity* are considered irrelevant. Threats to *construct validity* relates to not having the right operational measures for the concepts under study. The study uses a formal review protocol created using well-accepted guidelines for literature reviews [37, 38] and includes explicitly defined

---

[5] https://scholar.google.com.

Table 3. Quality assessment criteria and results

No.	Criteria	Possible responses	Results				
1	The research objective of this study is pertinent to the review	(0) No (1) Yes	0	1	Five studies eliminated		
			5	28			
			1	2	3	4	5
2	The study describes its context in sufficient detail	(1) Strongly Disagree (2) Disagree (3) Neither Agree/Disagree (4) Agree (5) Strongly Agree	6	2	8	10	2
3	The research design addresses study objectives		7	5	8	5	3
4	The research methods are described with adequate clarity		8	3	9	6	2
5	The findings & results lead to justifiable conclusions		6	3	7	9	3
6	The study's outcomes contribute to knowledge or practice		4	3	9	8	4
Distribution of mean scores across Criteria 2 to 5 (papers with mean scores < 2.5 eliminated)			6	1	13	6	2

data collection methods with clear inclusion/exclusion criteria and data extraction process. The authors individually validated this review protocol to help ensure conceptual relevance and mitigate potential threats to construct validity.

This systematic approach to the review enables reproducibility and enhances the reliability of the review. It also makes the review context very visible and makes the findings from selected studies amenable for generalization (or *external validity*). Having incorrect search methods, inappropriate search terms and time-spans, biases in data extraction and study selection, publication bias, and papers' inaccessibility are the leading causes for missing relevant primary studies [41]. Search terms are kept aligned to the research questions and selected based on agile software development and project portfolio management concepts. The search string is kept generic enough to include as many studies as possible that refer to the key terms of "agile" and "portfolio." The snowballing process and the searches across the six databases have helped minimize the risk of missing out on relevant studies.

Issues in the interpretation of data could lead to potential threats to the study's *conclusion validity*. An "audit trail" of review records maintained on an Excel spreadsheet capturing detailed reasons for including or excluding a study mitigates against threats of bias during data extraction. Studies varied in the detail provided around their methods, their settings, analyses performed, and the conclusions drawn. These are reflected in the quality assessment carried out in Stage 3, leading to the elimination of 12 studies (from the 33 studies available for quality assessment) due to inadequate rigor and detail, thus minimizing the risk of inaccuracy during data extraction.

4   Findings

The literature review identified 21 empirical studies relating to portfolio management in environments using agile software development practices (Source studies are listed in the Appendix and are referred to in upcoming sections using their identifiers ranging from S01 to S21). Figure 2 shows summarized bibliometric information.

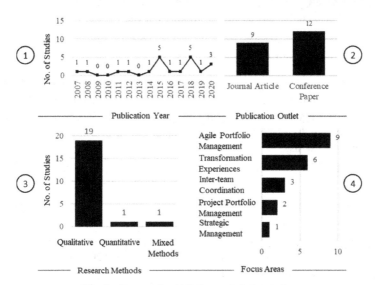

Fig. 2. Summarized bibliometric information

The distribution of publication dates (Fig. 2 – graph 1) reveals that research into this area is sporadic and that much of the limited work in this area appears in the last five years[6]. The selected studies included nine journal articles (43%) and 12 conference papers (57%) (Fig. 2 – graph 2).

Most studies (19 studies or 90%) were based on qualitative research designs (Fig. 2 – graph 3). The case study method was by far the most common approach used to explore

---

[6] Note that the search strategy had not used any date filters and the results include all available papers until early June 2020 when the search was conducted.

portfolio management in agile environments (17 studies or 80%), with the researcher(s) closely affiliated to the case organization(s).

The reviewed studies' focus areas were classified into five groups to appreciate the diversity of topics under research (Fig. 2 – graph 4). Nine studies (43%) investigated agile portfolio management practices; six studies (29%) described enterprise-level agile transformation efforts directly impacting portfolio management practices; three studies (14%) researched implications of inter-team coordination in agile environments; two studies (9%) explore project portfolio practices to enhance agility (although not directly referring to agile methods) and one (5%) that delves into areas of strategic management in context of portfolio management.

The studies called out potential effects of existing organizational mechanisms for managing IT investments (19 studies – 90%), human resources (15 studies – 71%), third-party vendor management (14 studies – 67%) and cultural aspects like acceptance by development teams and senior management commitment (21 studies – 100%) as crucial factors to consider while modifying portfolio processes. These general observations are similar to findings from past literature reviews on large-scale agile practices [32–35]. However, this review goes further to highlight a consistent need for portfolio processes to mature further to enable agility at the enterprise level.

4.1 Impacts on Portfolio Practices

In response to the research question RQ1, the review process studied the reported impacts on portfolio management practices and conceptually aggregated them into six key themes. These themes represent IT portfolio management aspects impacted by agile delivery methods and can be perceived as challenges in practice. Table 4 lists these impacted portfolio areas.

Table 4. Impacted portfolio areas

No.	Impacted portfolio areas	Source studies	Count
1	Portfolio strategic alignment	S01, S02, S03, S05, S06, S09, S10, S11, S13, S14, S16, S18, S19, S20, S21	15 (71%)
2	Continuous delivery	S03, S05, S07, S09, S13, S14, S17	7 (33%)
3	Adaptive nature	S01, S04, S07, S09, S12, S13, S20	7 (33%)
4	Learning through feedback	S03, S04, S07, S09, S14, S16, S17, S18,	8 (38%)
5	Financial processes	S01, S04, S06, S12, S13, S20	6 (29%)
6	Performance indicators	S09, S13, S15, S16, S18, S19, S20	7 (33%)

Portfolio Strategic Alignment. Agile teams are characterized by increased interactions within and across portfolio components and actors (like customers and stakeholders), which increases portfolio level complexities (S01, S05, S10, S18). Similarly, interdependencies and conflicts across multiple agile development teams are resolved through direct interactions across teams (S03, S21). Portfolio management practices need to evolve to keep these interactions aligned to the strategic business objectives (S18) and address project interdependencies (S11) within the portfolio.

While the emergent strategy can be supported through portfolio rebalancing or reconfiguration (S09, S11, S13, S14, S16, S19), some studies highlight applying a continuous improvement mindset to portfolio processes to enhance their capabilities to explore, sense, and respond to emergent strategy (S10, S18, S20). Portfolio processes should adequately communicate business strategy to all constituent teams (S01, S06, S18) to make dependencies visible, to create shared mental models to facilitate coordination (S02), and for planning project resourcing (S01, S09).

Continuous Delivery. Agile teams require ongoing portfolio prioritization and selection to maintain the constant cadence in delivering business outcomes through their backlogs for each cycle (S05, S09, S13, S17) and to better support inter-team coordination of dependencies (S14). This continuous portfolio process of "feeding the machine" (S03) maintains the overall project delivery schedule and release plans. Portfolio processes need streamlining and simplification to synchronize planning cycles across technical iterations and business (S16) to help agile teams obtain adequate backlog information just-in-time for upcoming delivery cycles (S07, S09, S13) and to avoid build-up of work items that could rapidly become obsolete over time (S05).

Adaptive Nature. The review observes a need for a leaner business case process (S01, S04, S07, S09, S12, S13, S20) to accommodate the adaptive and self-organizing nature of agile projects. Agile business cases provide "just enough" content needed to consider an IT investment option with details getting incorporated as requirements emerge with higher confidence (S12, S20) impact traditional portfolio governance and control processes relying on detailed business case assessment using project characteristics like scope, timelines, costs, benefits, and risks defined *a priori* [42]. Portfolio processes need to bridge gaps between existing organizational processes aligned with traditional stage-gate approaches and agile development processes (S04).

Learning Through Feedback. The classical portfolio management approach of measuring project outcomes against pre-defined success (or failure) criteria based on upfront plans is contrary to the agile way and can inhibit the organization from learning from its project experiences (S04, S18). Concepts like lean-startup and learning through experimentation (S17) in agile teams require portfolios to use continuous feedback mechanisms across the development lifecycle (S09) on projects constructed as proof-of-concept hypotheses (S04). Portfolio processes should extend the feedback-based learning mechanism from agile teams to adjacent business and management domains (S14) and sustain organizational learning (S03).

Traditional portfolio approaches assume resources to be fungible and continually (re)allocates them based on business priorities resulting in frequent context switching that can create unrest (S14) and limit learning ability (S16).

Financial Processes. Portfolio mechanisms need to bridge the gap between shorter and adaptive planning cycles required for agile development with the longer horizons and stable plans mandated by the business (S04, S06, S12, S13). Traditional project valuation methods (using measures like Net Present Value and Earned Value Analysis) do not adequately support the use of agile value metrics (like Net Promoter Score, product demo feedbacks, or metrics like cycle-time and throughput). (S01, S04) The evolving business case process also reflects this need to raise funding to a level higher than an individual project (S20).

Performance Indicators. Portfolio metrics need to reflect enterprise performance at the highest level to reflect the business impact of projects implemented (S09), and not just be considered output control mechanisms (S13). Many of the traditional metrics, like the Schedule Performance Index (SPI) and the Cost Performance Index (CPI), have no relevance in an agile environment, requiring portfolio management to identify more insightful metrics (S20). The sole study focusing on reporting in agile portfolios describes it as an "information exchange mechanisms across boundaries of knowledge domains" (S15). The portfolio could institute appropriate structures and routines (like a PMO) to coordinate this knowledge exchange (S16, S19) and periodically assess how well the IT portfolio and its constituent projects adapt to environmental changes (S18).

4.2 Agility Approaches in Practice

The review identified various portfolio practices that enterprises adopt to address (or at least minimize) impacts from agile development methods to address the research question RQ2. The respective organizational context plays a role in adopting these practices. Since most studies in the review had an exploratory or descriptive focus, they do not provide any causal insights into how a specific practice contributes to portfolio agility. Table 5 shows these practices mapped to their respective portfolio areas of impact.

Portfolio Strategic Alignment. Portfolio backlogs showing strategic investment themes and how they relate to portfolio components like epics, features, and stories (S09, S11), often implemented as Kanban (S01) or portfolio walls (S09), provide end-to-end portfolio visibility to enterprise stakeholders. They allow portfolios to continuously adapt to upstream changes in business strategy or product line directions and ensure appropriate downstream adaptations within teams (S05). End-to-end portfolio visibility also aids in streamlining coordination across teams (S06, S11, S16, S21), strengthens the communication process, facilitates joint decision-making, builds trust, and enhances collaboration within the team and across stakeholders (S04, S13, S14). The PMO is an enabling structure to manage this visibility (S14, S19, S20).

Continuous Delivery. Agile portfolios advocate short portfolio cycles (S10, S20) synchronized at multiple integration points (S06) with project approvals and epic/solution details provided just-in-time for immediately upcoming cycles (S01, S03, S05, S09) to ensure adequate utilization of the development pipeline and to avoid requirements or projected benefits becoming stale while in the development pipeline. Collaborative and visual planning led to better continuous planning outcomes (S11, S17). Continuous prioritization of the portfolio (S04, S09) based on ongoing feedback keeps development teams aligned to portfolio objectives (S13).

Table 5. Portfolio level approaches adopted to support agile methods

No.	Portfolio area	Approach adopted	Source studies	Count
1	Portfolio Strategic Alignment	Portfolio backlogs provide end-to-end visibility	S01, S02, S04, S06, S07, S09, S10, S11, S13, S14, S16, S19, S20, S21	14 (67%)
		PMO structures to facilitate visibility	S14, S19, S20	3 (14%)
2	Continuous Delivery	Shorter portfolio cycles	S06, S10, S20	3 (14%)
		Collaborative planning	S11, S17	2 (10%)
		JIT approvals	S01, S03, S05, S09	4 (19%)
		Continuous Prioritization	S04, S09, S13	3 (14%)
3	Adaptive nature	Customer value as the basis for evaluation	S01, S06, S09, S13	4 (19%)
		Shorter planning cycles	S10, S20	2 (10%)
		Planning at higher levels of abstraction	S04, S09, S14, S20	4 (19%)
4	Learning through feedback	None	None	
5	Financial Processes	Continuous forecasts	S09, S12, S13, S20	4 (19%)
6	Performance indicators	PMO structures to facilitate reporting	S13, S15, S19	3 (14%)

Adaptive Nature. Business case evaluation and prioritization utilize portfolio parameters based on customer value (S01, S06, S09, S13), although portfolio practitioners have found it difficult to evolve acceptable, consistent, and measurable definitions of *"value"* (S11). Traditional portfolio practices can support the adaptive nature of agile methods by having shorter portfolio cycles (S06, S10, S20) and by defining projects as features or value propositions (S04) at higher levels of abstraction (S20).

Learning Through Feedback. While agile methods applied at the team level facilitates learning through feedback cycles, none of the studies reported any conscious portfolio practice to facilitate portfolio level learning. One study recommended knowledge replication as a potential practice (S07) but did not offer any further detail.

Financial Processes. Agile organizations are moving from budget controls to more of an emergent outcome control model (S13). Rolling wave forecasts, where a continuous cadence of forecasts replaces the traditional fixed horizon budget process, is a significant shift in the way enterprises manage project funding (S09, S12, S20). Another meaningful change is the shift towards funding product/feature teams instead of projects (S12) and moving cost center planning to a more aggregated level (S11).

Performance Indicators. One study identifies a set of reporting practices used in portfolios to share information across knowledge domain boundaries (S15) effectively. PMOs have a role in consolidating and disseminating metrics across the portfolio, especially end-to-end metrics like "Time to Market Improvement" and "Customer Satisfaction" (S13, S15, S19).

5   Discussion

This literature review identifies six portfolio management aspects impacted by incorporating agile efforts in the portfolio (Sect. 4.1) to answer the first research question (RQ1). The various antecedents linked to the challenges identified provide the background to frame further research studies into these areas. In response to the second research question (RQ2), the review recognizes practices/approaches adopted by enterprises to achieve agility in their IT portfolios (Sect. 4.2). These practices/approaches are mapped to the six aspects impacted by agile projects (RQ1) to reflect potential resolutions to the challenges identified. This review does not attempt to evaluate their relative merits due to the varying level of exploratory detail across studies. Further empirical evaluation of these practices and their contribution to portfolio success and agility through each of the six portfolio management aspects is recommended.

Many studies in this review (S01, S05, S06, S11, S13, S14, S18) have observed that the field of agile portfolio management is relatively unexplored. This literature survey shares the same view based on the low number of empirical studies (21 studies) identified. An analysis of overall scores[7] from the quality assessment conducted in Stage 3 leads to an inference around the relatively low strength of evidence across many studies, possibly requiring further research to validate their theoretical contribution claims.

5.1   Implications of Findings

Practitioners involved in scaling agile practices across an enterprise should view the six portfolio management aspects impacted by agile delivery efforts as crucial factors in enabling portfolio level agility. Enterprises should reflect on the identified practices/approaches using their respective organizational context since the reviewed literature does not identify any specific causal relationship.

There are three implications to research and practice from the findings in this review.

1. *Shifting from reactive to proactive approaches.* Portfolio management literature [10, 43, 44] acknowledges that maintaining strategic alignment is one of portfolio management's key objectives, achieved through top-down approaches aligning

---

[7] Some descriptive statistics of the overall quality assessment scores are as follows: n = 21, mean = 3.50, median = 3.20, min = 2.6, max = 4.8.

enterprise objectives to IT priorities [11]. This traditional approach towards portfolio management appears reactive as it focuses on reconfiguring its components as a response to business strategy changes. Studies in this review describe portfolio backlogs and Kanban as tools to provide visibility into portfolio components, their alignment to strategic themes, and their priorities (S01, S09, S11), allowing for an effective response to changes, once sensed. It is unclear how these tools help portfolios become proactive in sensing changes to their dynamic environments to enable enterprise agility.

One of the reviewed studies (S11) indicates that it could take months before a new project is accepted into a supposedly agile portfolio – inhibiting the ability to enable continuous delivery. Two studies (S06, S18) offer recommendations around continuous portfolio exploration, and further investigation is needed to explain how portfolios could shift to more proactive approaches to support continuous delivery expectations.

2. **Adopting a systems-thinking approach.** Sweetman et al. (S18) present a unique view of an agile portfolio as a complex adaptive system. Cao et al. [45] had proposed the study of agile software development projects as a dynamic, integrated system, given its use of autonomous teams, frequent iterations incorporating feedback, and continuous adaptation of product features. Therefore, a portfolio system, characterized by its routines, structures, and values (S14), essentially becomes a "system of systems" consisting of various individual agile efforts.

A systems-thinking approach could explore a portfolio system as a set of interactions across multiple interconnected and interdependent components to collectively achieve the portfolio objectives.

3. **Changes to existing organizational processes.** Studies indicate that existing strategic planning and investment management processes impact agile portfolio implementations (S09, S12, S13, S20). Cao et al. [46] suggest that agile efforts require modified enterprise project budgeting structures and processes due to limitations of traditional project appraisal, expense capitalization, and contract valuation methods [47]. Beyond Budgeting [48], Multi-Level Budgeting [49], and Real Options [50] are alternate options to be further explored. Krebs [24] recommends the use of dynamic financial models to drive portfolio agility, while the Scaled Agile Framework (SAFe) advocates the practice of lean portfolio management [25], applying principles from lean systems [51] to align strategy and execution. Dikert et al. [34] recognize the crucial role of non-IT functions in successfully scaling agile practices across the enterprise and recommends further research into this area.

5.2 Limitations of This Study

The explicitly defined review protocol detailing the various stages of the process mitigates most limitations related to potential biases in study selection and data extraction. The more experienced researchers independently validated this review protocol to reduce bias in the process. Although a sole researcher conducted the multi-stage study selection process due to resource constraints, the "audit trail" of inclusion/exclusion decisions helped traceability while reviewing the work.

Despite a widened search process to accommodate as many studies as possible across the field of inquiry, only a few empirical studies (21 studies) were identified, meeting all the pre-defined selection criteria. Coupled with the relatively low scores observed in the quality assessment process, this indicates a need for further empirical research in this area.

6   Conclusions and Future Research Directions

The six aspects of portfolio management identified in this review as impacted by agile delivery (in response to RQ1) and the various solution approaches described (in response to RQ2) present opportunities for future exploration to identify causal explanations, configurational patterns, and the nature/extent of their relationships to agility and portfolio success. While some of the reviewed studies illustrated how portfolios are reconfigured to 'respond' to changes, there was no depiction of how portfolios 'sense' changes or 'learn' from these changes to optimize future responses. Future studies are needed to understand the "sensing' and 'learning' aspects of portfolio agility.

Using a systems-thinking lens to model and diagnose agile portfolio structures, processes, and interactions is another potential research avenue, leading to the definition and analysis of possible portfolio methods enabling agility at different levels of the organization. Another research direction for the future could be around the systemic interfaces and dependencies of adjacent organizational processes (like HR and Finance) on portfolio practices and their impacts on agility.

This literature review makes three contributions to academic knowledge. Firstly, it synthesizes current knowledge of how project portfolios enable IT agility while incorporating agile delivery efforts. Secondly, it responds to the specific questions by identifying six portfolio management aspects impacted by agile delivery practices and a set of current practices used within enterprises to contribute to portfolio agility. Finally, the implications of these findings have helped identify possible future research directions, some of which are explored by the authors in the upcoming stages of designing an agile portfolio management framework.

Appendix – Selected Studies

ID	Citation
S01	Ahmad, M.O., Lwakatare, L.E., Kuvaja, P., Oivo, M., Markkula, J.: An empirical study of portfolio management and Kanban in agile and lean software companies. Journal Of Software: Evolution and Process 29(6), 1-16 (2017)
S02	Bjørnson, F.O., Wijnmaalen, J., Stettina, C.J., Dingsøyr, T.: Inter-team coordination in large-scale agile development: A case study of three enabling mechanisms. In: International Conference on Agile Software Development 2018, pp. 216-231. Springer (2018)

(continued)

(continued)

ID	Citation
S03	Dingsøyr, T., Moe, N.B., Fægri, T.E., Seim, E.A.: Exploring software development at the very large-scale: a revelatory case study and research agenda for agile method adaptation. Empirical Software Engineering 23(1), 490-520 (2018)
S04	Hansen, L.K., Brandt, C.J., Svejvig, P., Kampf, C.E.: Agile project portfolio management, new solutions and new challenges: findings from four agile organizations. In: EURAM Conference (2020)
S05	Hoffmann, D., Ahlemann, F., Reining, S.: Reconciling alignment, efficiency, and agility in IT project portfolio management: Recommendations based on a revelatory case study. International Journal of Project Management 38(2), 124-136 (2020)
S06	Horlach, B., Schirmer, I., Drews, P.: Agile portfolio management: Design goals and principles. In: 27th European Conference on Information Systems (ECIS), Stockholm-Uppsala, Sweden 2019. AIS Electronic Library (AISeL) (2019)
S07	Imbrizi, F.G., Maccari, E.A.: Agile Software Development and Project Portfolio Management in Dynamic Environments: An exploratory case study. In: International Association for Management of Technology (2014)
S08	Kaufmann, C., Kock, A., Gemünden, H.G.: Emerging strategy recognition in agile portfolios. International Journal of Project Management (2020)
S09	Laanti, M., Sirkiä, R., Kangas, M.: Agile Portfolio Management at Finnish Broadcasting Company Yle. In: Scientific Workshop Proceedings of the XP2015, pp. 1-7. ACM (2015)
S10	Petit, Y.: Project portfolios in dynamic environments: Organizing for uncertainty. International Journal of Project Management 30(5), 539-553 (2012)
S11	Rautiainen, K., Von Schantz, J., Vähäniitty, J.: Supporting scaling agile with portfolio management: Case Paf.com. In: 44th Hawaii International Conference on System Sciences 2011, pp. 1-10. IEEE (2011)
S12	Sirkiä, R., Laanti, M.: Adaptive Finance and Control: Combining Lean, Agile, and Beyond Budgeting for Financial and Organizational Flexibility. In: 48th Hawaii International Conference on System Sciences 2015, pp. 5030-5037 (2015)
S13	Smeekes, I., Borgman, H., Heier, H.: A Wheelbarrow Full of Frogs: Understanding Portfolio Management for Agile Projects. In: 51st Hawaii International Conference on System Sciences 2018, pp. 5473-5482. IEEE (2018)
S14	Stettina, C.J., Horz, J.: Agile portfolio management: An empirical perspective on the practice in use. International Journal of Project Management 33(1), 140-152 (2015)
S15	Stettina, C.J., Schoemaker, L.: Reporting in agile portfolio management: Routines, metrics and artefacts to maintain an effective oversight. In: International Conference on Agile Software Development 2018, pp. 199-215 (2018)
S16	Stettina, C.J., Smit, M.N.W.: Team portfolio scrum: An action research on multitasking in multi-project scrum teams. In: International Conference on Agile Software Development 2016, pp. 79-91 (2016)

(continued)

(*continued*)

ID	Citation
S17	Suomalainen, T., Kuusela, R., Tihinen, M.: Continuous planning: an important aspect of agile and lean development International Journal of Agile Systems and Management 8(2), 132-162 (2015)
S18	Sweetman, R., Conboy, K.: Portfolios of Agile Projects A Complex Adaptive Systems' Agent Perspective. Project Management Journal 49(6), 18-38 (2018)
S19	Tengshe, A., Noble, S.: Establishing the Agile PMO: Managing variability across Projects and Portfolios. In: Proceedings of Agile 2007, pp. 188-193. IEEE (2007)
S20	Thomas, J.C., Baker, S.W.: Establishing an agile portfolio to align IT investments with business needs. In: Proceedings of Agile 2008, pp. 252-258. IEEE (2008)
S21	Vlietland, J., van Vliet, H.: Towards a governance framework for chains of Scrum teams. Information and Software Technology 57, 52-65 (2015)

References

1. Weill, P., Woerner, S.L.: Thriving in an increasingly digital ecosystem. MIT Sloan Manag. Rev. **56**(4), 27 (2015)
2. Overby, E., Bharadwaj, A., Sambamurthy, V.: Enterprise agility and the enabling role of information technology. Eur. J. Inf. Syst. **15**(2), 120–131 (2006)
3. Leonhardt, D., Haffke, I., Kranz, J., Benlian, A.: Reinventing the IT function: the role of IT agility and IT ambidexterity in supporting digital business transformation. In: Proceedings of the 25th European Conference on Information Systems (ECIS), pp. 968–984 (2017)
4. Serrador, P., Pinto, J.K.: Does Agile work?—A quantitative analysis of agile project success. Int. J. Project Manag. **33**(5), 1040–1051 (2015)
5. Dingsøyr, T., Moe, N.B.: Towards principles of large-scale agile development. In: Dingsøyr, T., Moe, N.B., Tonelli, R., Counsell, S., Gencel, C., Petersen, K. (eds.) XP 2014. LNBIP, vol. 199, pp. 1–8. Springer, Cham (2014). https://doi.org/10.1007/978-3-319-14358-3_1
6. Horlach, B., Böhmann, T., Schirmer, I., Drews, P.: IT governance in scaling agile frameworks. In: Proceedings of the Multikonferenz Wirtschaftsinformatik, Lüneburg 2018, pp. 1789–1800 (2018)
7. Levine, H.A.: Project Portfolio Management: A Practical Guide to Selecting Projects, Managing Portfolios, and Maximizing Benefits. Jossey-Bass, San Francisco (2005)
8. Meskendahl, S.: The influence of business strategy on project portfolio management and its success—a conceptual framework. Int. J. Project Manag. **28**(8), 807–817 (2010)
9. Blichfeldt, B.S., Eskerod, P.: Project portfolio management–there's more to it than what management enacts. Int. J. Project Manag. **26**(4), 357–365 (2008)
10. Cooper, R.G., Edgett, S.J., Kleinschmidt, E.J.: New problems, new solutions: making portfolio management more effective. Res.-Technol. Manag. **43**(2), 18–33 (2000)
11. Müller, R., Martinsuo, M., Blomquist, T.: Project portfolio control and portfolio management performance in different contexts. Proj. Manag. J. **39**(3), 28–42 (2008)
12. McFarlan, F.W.: Portfolio approach to information systems. Harvard Bus. Rev. **59**(5), 9 (1981)
13. Markowitz, H.: Portfolio selection. J. Finance **7**(1), 15 (1952)
14. US General Accounting Office (GAO): Improving Mission Performance through Strategic Information Management and Technology (1994). https://www.gao.gov/special.pubs/ai94115.pdf

15. Maizlish, B., Handler, R.: IT (Information Technology) Portfolio Management Step-by-step: Unlocking the Business Value of Technology. Wiley, Hoboken (2005)
16. De Haes, S., Van Grembergen, W.: An exploratory study into IT governance implementations and its impact on business/IT alignment. Inf. Syst. Manag. **26**(2), 123–137 (2009)
17. Ahola, T., Ruuska, I., Artto, K., Kujala, J.: What is project governance and what are its origins? Int. J. Project Manag. **32**(8), 1321–1332 (2014)
18. Kujala, J., Aaltonen, K., Gotcheva, N., Pekuri, A.: Key dimensions of project network governance and implications to safety in nuclear industry projects. In: EURAM 2016: Manageable Cooperation? (2016)
19. Digital.ai: The 14th Annual State of Agile Survey (2020). https://explore.digital.ai/state-of-agile/14th-annual-state-of-agile-report
20. Sutherland, J., Schwaber, K.: The Scrum Guide. scrumguides.org (2017). https://www.scrumguides.org/scrum-guide-2017.html
21. Beck, K., et al.: Manifesto for Agile Software Development (2001). https://agilemanifesto.org/
22. Nerur, S., Balijepally, V.: Theoretical reflections on agile development methodologies. Commun. ACM **50**(3), 79–83 (2007)
23. Budzier, A., Flyvbjerg, B.: Making sense of the impact and importance of outliers in project management through the use of power laws. In: Proceedings of IRNOP (International Research Network on Organizing by Projects), Oslo (2013)
24. Krebs, J.: Agile Portfolio Management. Microsoft Press, Redmond (2008)
25. Scaled Agile Framework: SAFe 5.0 Framework - SAFe Big Picture. Scaled Agile Framework (2020). https://www.scaledagileframework.com/
26. Hansen, L.K., Kræmmergard, P.: Discourses and theoretical assumptions in IT project portfolio management: a review of the literature. Int. J. Inf. Technol. Proj. Manag. (IJITPM) **5**(3), 39–66 (2014)
27. Hoda, R., Kruchten, P., Noble, J., Marshall, S.: Agility in context. In: 2010 Proceedings of the ACM International Conference on Object Oriented Programming Systems Languages and Applications, pp. 74–88. ACM (2010)
28. Karlström, D., Runeson, P.: Integrating agile software development into stage-gate managed product development. Empir. Softw. Eng. **11**(2), 203–225 (2006). https://doi.org/10.1007/s10664-006-6402-8
29. Abrantes, R., Figueiredo, J.: Resource management process framework for dynamic NPD portfolios. Int. J. Project Manag. **33**(6), 1274–1288 (2015)
30. Lee, G., Xia, W.: Toward agile: an integrated analysis of quantitative and qualitative field data on software development agility. MIS Q. **34**(1), 87–114 (2010)
31. Theobald, S., Schmitt, A., Diebold, P.: Comparing scaling agile frameworks based on underlying practices. In: Hoda, R. (ed.) XP 2019. LNBIP, vol. 364, pp. 88–96. Springer, Cham (2019). https://doi.org/10.1007/978-3-030-30126-2_11
32. Alqudah, M., Razali, R.: A review of scaling agile methods in large software development. Int. J. Adv. Sci. Eng. Inf. Technol. **6**(6), 828–837 (2016)
33. Lappi, T., Karvonen, T., Lwakatare, L.E., Aaltonen, K., Kuvaja, P.: Toward an improved understanding of agile project governance: a systematic literature review. Proj. Manag. J. **49**(6), 39–63 (2018)
34. Dikert, K., Paasivaara, M., Lassenius, C.: Challenges and success factors for large-scale agile transformations: a systematic literature review. J. Syst. Softw. **119**, 87–108 (2016)
35. Ahmad, M.O., Dennehy, D., Conboy, K., Oivo, M.: Kanban in software engineering: a systematic mapping study. J. Syst. Softw. **137**, 96–113 (2018)
36. Tranfield, D., Denyer, D., Smart, P.: Towards a methodology for developing evidence-informed management knowledge by means of systematic review. Br. J. Manag. **14**(3), 207–222 (2003)

37. Kitchenham, B., Charters, S.: Guidelines for performing systematic literature reviews in software engineering (2007)
38. Okoli, C.: A guide to conducting a standalone systematic literature review. Commun. Assoc. Inf. Syst. **37**(1), 43 (2015)
39. Wohlin, C.: Guidelines for snowballing in systematic literature studies and a replication in software engineering. In: 2014 Proceedings of the 18th International Conference on Evaluation and Assessment in Software Engineering, pp. 1–10 (2014)
40. Cruzes, D.S., Dyba, T.: Recommended steps for thematic synthesis in software engineering. In: 2011 International Symposium on Empirical Software Engineering and Measurement, pp. 275–284. IEEE (2011)
41. Zhou, X., Jin, Y., Zhang, H., Li, S., Huang, X.: A map of threats to validity of systematic literature reviews in software engineering. In: 2016 23rd Asia-Pacific Software Engineering Conference (APSEC), pp. 153–160. IEEE (2016)
42. Archer, N.P., Ghasemzadeh, F.: An integrated framework for project portfolio selection. Int. J. Project Manag. **17**(4), 207–216 (1999)
43. Killen, C.P., Hunt, R.A., Kleinschmidt, E.J.: Managing the new product development project portfolio: a review of the literature and empirical evidence. In: PICMET 2007 Portland International Conference on Management of Engineering & Technology 2007, pp. 1864–1874. IEEE (2007)
44. Martinsuo, M., Lehtonen, P.: Role of single-project management in achieving portfolio management efficiency. Int. J. Project Manag. **25**(1), 56–65 (2007)
45. Cao, L., Ramesh, B., Abdel-Hamid, T.: Modeling dynamics in agile software development. ACM Trans. Manag. Inf. Syst. (TMIS) **1**(1), 1–26 (2010)
46. Cao, L., Mohan, K., Ramesh, B., Sarkar, S.: Adapting funding processes for agile IT projects: an empirical investigation. Eur. J. Inf. Syst. **22**(2), 191–205 (2013)
47. Moran, A.: Managing Agile: Strategy, Implementation, Organisation and People. Springer, Cham (2015). https://doi.org/10.1007/978-3-319-16262-1
48. Sahota, M., Bogsnes, B., Nyfjord, J., Hesselberg, J., Drugovic, A.: Beyond Budgeting: a Proven Governance System Compatible with Agile Culture. BBRT (2014). http://bbrt.co.uk/bbfiles/BeyondBudgetingAgileWhitePaper_2014.pdf
49. Knaster, R., Leffingwell, D.: SAFe 4.5 Distilled: Applying the Scaled Agile Framework for Lean Enterprises. Addison-Wesley Professional, Boston (2018)
50. Racheva, Z., Daneva, M.: Using measurements to support real-option thinking in agile software development. In: Proceedings of the 2008 International Workshop on Scrutinizing Agile Practice (Shoot-out at the agile corral), pp. 15–18. ACM (2008)
51. Reinertsten, D.G.: The Principles of Product Development Flow: Second Generation Lean Product Development. Celeritas, Redondo Beach (2009)

A Systematic Literature Review on Implementing Non-functional Requirements in Agile Software Development: Issues and Facilitating Practices

Aleksander Jarzębowicz$^{(\boxtimes)}$ and Paweł Weichbroth

Gdańsk University of Technology, Faculty of Electronics,
Telecommunications and Informatics, Department of Software Engineering,
11/12 Gabriela Narutowicza Street, 80-233 Gdańsk, Poland
{aleksander.jarzebowicz,pawel.weichbroth}@pg.edu.pl
http://www.pg.edu.pl

Abstract. Agile Software Development methods have become a widespread approach used by the software industry. Non-functional requirements (NFRs) are often reported to be a problematic issue for such methods. We aimed to identify (within the context of Agile projects): (1) the issues (challenges and problems) reported as affecting the implementation of NFRs; and (2) practices that facilitate the successful implementation of NFRs. We conducted a systematic literature review and processed its results to obtain a comprehensive summary. We were able to present two lists, dedicated to issues and practices, respectively. Most items from both lists, but not all, are related to the requirements engineering area. We found out that the issues reported are mostly related to the common themes of: NFR documentation techniques, NFR traceability, elicitation and communication activities. The facilitating practices mostly cover similar topics and the recommendation is to start focusing on NFRs early in the project.

Keywords: Non-functional Requirements · Quality requirements · Agile Software Development · Agile requirements engineering · Systematic literature review

1 Introduction

Agile Software Development (ASD) is an iterative approach to delivering software products. The term "agility" implies adaptability [1], flexibility [2], and close collaboration with the customer [3]. An Agile approach assumes sensible values such as trust [4], responsibility [5] and loyalty [6]. Around half of organizations have now been applying Agile practices for over three years to adopt change and transformation management [7]. Moreover, the results from a survey conducted

© Springer Nature Switzerland AG 2021
A. Przybyłek et al. (Eds.): LASD 2021, LNBIP 408, pp. 91–110, 2021.
https://doi.org/10.1007/978-3-030-67084-9_6

in 2018 among software industry practitioners show that 97% of respondents declared using Agile methods [8]. In fact, the benefits of adopting Agile practices have been reported in many studies [9–11], indicating an increase of team productivity, motivation and discipline, as well as overall software quality, just to name a few.

Indeed, software quality is an important aspect to be considered during the software lifecycle [12,13], usually defined in terms of high-level attributes [14]. Alternatively, one can impose additional constraints on the behavior of the system. In other words, the required properties (attributes and constraints) are specified as non-functional requirements (hereafter, NFRs), in addition to functional requirements (FRs). Since the beginning of software development as a job role, NFRs have been recognized as critical factors that affect the acceptance and use of the products by the users [15].

In fact, to mitigate the risk of users' dissatisfaction by misunderstanding or disregarding their expectations and needs, active user involvement is imperative in ASD [16–18]. However, one question arises naturally: Does this user engagement bring other risks, and does the development team need to find a balance between risk and benefits?

Undeniably, the search for the answer to this question has been the subject of vast research [19–21], since the introduction of the Agile Manifesto [22]. Nevertheless, few studies provide an evidence-based review and analysis on the subject of implementing NFRs, in particular regarding the issues that could arise with the advance of ASD, as well as the practices that have been documented as successful facilitators.

The values and principles followed in ASD also result in practices different than those used in more traditional software development methods. It includes requirements engineering practices [23], which e.g. assume continuous close cooperation with the customer [24], put more emphasis on face-to-face communication [25], and use less formal techniques like collaborative games [26].

Both researchers and practitioners have repeatedly noted the challenges in Agile requirements engineering. For example, the results from a Delphi study [27], performed in 2017 in a group of 26 experts, show that one of the recognized challenges is to "establish non-functional requirements", which has been reported by prior other studies [25,28,29]. The comprehensive know-how with regard to the more detailed challenges and relevant counteractions is not available though. The only available secondary study focusing on NFRs in ASD at the time we started our research was the SLR by Alsaqaf et al. [30]. That SLR was considered by its authors as a starting point for further empirical studies and several primary studies were published since then. To systematize the current state of the art, in this paper, we put forward these two following research questions (RQs):

1. What issues affect the identification and implementation of non-functional requirements in ASD?
2. What practices facilitate the successful identification and implementation of non-functional requirements in ASD?

Therefore, the goal of this study is to review and analyze the existing studies and their outcomes and to summarize the documented issues and applied practices, in the extent of NFR identification and implementation, within the ASD context. To provide evidence-based and state-of-the-art answers to the above questions we conducted a systematic literature review (SLR).

By design, the results of this study are complementary to the existing body of knowledge by providing the following contributions to the software engineering discipline: the collections of (i) the current issues (challenges and problems), and (ii) the explicit practices that, respectively, affect and facilitate the identification and implementation of NFRs within ASD. Moreover, the findings in this paper entail useful implications for researchers and practitioners alike. In this context, while the former group might be interested in investigating the impact of particular issues on the success (failure) of ASD projects, the latter group might be willing to mitigate those issues by adopting the practices in the scope and content due to the current needs and priorities.

The remainder of this paper is laid out as follows. Section 2 describes the rationale behind implementing NFRs. Section 3 provides the description of the research methodology, applied to conduct the systematic literature review. The results are given in Sect. 4, followed by their discussion in Sect. 5. Finally, the paper is concluded in Sect. 6.

2 Rationale Behind Implementing NFRs

Generally speaking, non-functional requirements (NFRs), also known as quality requirements, define the users' expectations and needs regarding a software product, as well as their particular notions of its qualities. According to Svensson et al. [31], the most important quality attributes in industrial practice relate to usability, performance, reliability, stability, safety, security/integrity, compliance, maintainability, reusability and interoperability. Unmistakably, NFRs have great importance in software product development [31–33].

Besides this, NFRs can also impose global constraints on a software product [34], arising from all of its parts as well as from interdependencies between them [35]. In other words, NFRs put constraints on how the product's functions must work [36]. Overlooking or even neglecting information related to quality facets negatively affects the final product. Ironically, although it might be surprisingly different from common sense, NFR-related errors are still claimed to be the most difficult to correct, and the most expensive [37]. It is a major risk, especially considering that in recent years software defects have become the dominant cause of user outage [38].

Undeniably, both researchers and practitioners from ASD communities have seen the need to capture, document and prioritise NFRs [39]. For instance, Microsoft, the largest software and programming company worldwide [40], recommends capturing functional and non-functional requirements alike, since the former indicate whether the application does the right thing, while the latter determine whether the application does those things well [41]. Oracle, the second largest software corporation, argues that "the key to successful software development is that all stakeholders develop a clear and uniform understanding of application requirements" [42].

Furthermore, we also acknowledge the importance of NFRs as the major external quality facets of the software products from the user's perspective [43]. The questions addressed in this study are narrowed to ASD, which assumes having the user(s) actively involved. If one compares Agile with traditional approaches, this involvement is not limited to the early stages of the development process. On the contrary, Agile development principles encourage active user involvement, being generally considered to contributing to user satisfaction [44, 45] and project success [46].

3   Methodology

We designed and executed the systematic literature review following the guidelines for SLR studies in software engineering elaborated by Kitchenham and Charters [47]. The definition of the search query and query execution in Scopus (phase 1 of SLR process) are shared with our other study aimed at identification of particular NFR-related requirements engineering techniques [48]. The inclusion/exclusion criteria were however defined with respect to this study's aim and subsequent phases of the SLR process were conducted separately in each of two studies.

We chose to rely on a single publication database (Elsevier Scopus). Scopus was selected because it indexes a large number of journals and conferences [49] and enables a single search query to access items from a broad variety of publishers [50]. It is worth noting here that in several other SLR studies similar to ours (e.g. [30, 51]) similar strategies were applied, in particular exclusively relying on the Scopus database.

3.1   Inclusion and Exclusion Criteria

The papers were eligible based on the five following inclusion criteria:

- peer-reviewed papers (I1);
- papers in English (I2);
- papers published since 2008 (I3);
- papers related to the software engineering domain (I4);
- papers covering Agile development and NFRs (I5).

The papers were screened prior to acceptance and were further rejected if they had any of the following exclusion criteria:

- papers not providing any information about NFR issues or practices in ASD (E1);
- papers not available for download, despite extensive search (E2);
- papers reporting the same results covered by another source included in SLR - in such cases the latest paper was included (E3);

– papers dedicated to a very specific subarea of NFRs (e.g. with proposals of advanced methods of establishing security requirements) (E4).

We focused on papers published since 2008 to include all works published in the last 12 years before the conduction of the SLR. We also decided to include papers dedicated to a specific project context (e.g. large-scale distributed development), but to exclude papers with very narrow scope (E4) e.g. with advanced dedicated analysis methods suggested as facilitating practices for security requirements, which are hard to consider as an issue or practice regarding the whole category of NFRs.

3.2 Search Query Definition

As we had performed some initial searches before planning the SLR, we were aware that sources dedicated to this topic of interest are rather scarce. This led us to the decision to cast a wider net and try to identify all sources focusing on NFRs in Agile, thus we used more generic keywords instead of those exactly matching our RQs (e.g. "challenges" or "practices").

The following search string was used:

TITLE-ABS-KEY ((agile OR scrum OR lean OR xp OR kanban) AND (nfr OR "non-functional requirements" OR "quality requirements")) AND PUBYEAR > 2007 AND (LIMIT-TO(DOCTYPE, "cp") OR LIMIT-TO (DOCTYPE, "ar") OR LIMIT-TO (DOCTYPE, "ch")) AND (LIMIT-TO (SUBJAREA, "COMP") OR LIMIT-TO (SUBJAREA, "ENGI") OR LIMIT-TO (SUBJAREA, "MATH") OR LIMIT-TO (SUBJAREA, "BUSI") OR LIMIT-TO (SUBJAREA, "DECI")) AND (LIMIT-TO (LANGUAGE, "English"))

The string includes various methods that could possibly be mentioned in the title, keywords etc. instead of the generic "Agile" term. We also provided alternative terms commonly used to denote an NFR. The types of documents mentioned in the search string match peer-reviewed papers. The specification of subject areas resulted from our knowledge, in particular on how some sources (especially the series that include conference proceedings as its volumes) are classified and indexed by Scopus. The search in titles, abstracts and keywords was chosen as the most comprehensive option available (Scopus does not enable searching the contents of full texts).

3.3 Search Strategy

We defined a process that comprised 3 main phases:

1. Execution of the search query.
2. Manual review of titles, keywords and abstracts of the papers retrieved from the search to exclude those not related to the topic of NFR in an Agile context.
3. Manual review of each remaining paper's full text in order to decide whether to finally include it or not. Identification of information pieces relevant to our RQs and assigning codes to them.

3.4   Search Execution

The results of the 3 phases defined in the previous section were as follows:

Phase 1: The search was executed on November 28th 2019. Despite including several alternative keywords in the search string, the search returned only 159 papers. This confirmed our initial suspicions that the topic of "NFR in Agile" is not widely addressed in the scientific papers, at least those indexed by Scopus.

Phase 2: The results retrieved by the Scopus search engine (that include title, keywords and the abstract of each paper found) were manually reviewed. It allowed us to verify the findings against I5 criterion more precisely than in the case of relying on an automated search and to reject papers that reported nothing on NFRs (for example, several papers referring to "quality requirements" turned out to interpret this term as "well-documented/valid requirements" instead of "requirements regarding system quality"). As a result, 71 papers were retained at the end of this phase.

Phase 3: In this phase, the papers were reviewed and checked against exclusion criteria E1-E4. Finally, 44 papers were qualified to extract information. Moreover, we evaluated the papers with regard to *(i)* the use of appropriate and rigorous research methods, *(ii)* clarity and coherence of the research findings, and *(iii)* providing a validation of the proposed approach. During the review, apart from just deciding on the paper's final classification, the fragments relevant to the RQs were identified and provided with codes to summarize the findings. Next, the codes were reviewed to identify similarities, and related codes were grouped into the more generic ones presented in the Results section.

4   Results

The final results of the SLR are presented in Tables 1 and 2. For each issue reported and facilitating practice suggested, a list of papers mentioning it is provided ("Sources" column). We also explicitly distinguish issues/practices related to requirements engineering activities ("RE" column) from those that should rather be associated with e.g. testing, architectural design or project management. Both tables are sorted starting from with the items quoted by the most sources. The elaboration of results with respect to the answers they provide to RQs is provided in 4.1 and 4.2.

4.1   What Issues Affect the Identification and Implementation of Non-functional Requirements in ASD?

The most frequently reported issue concerns neglecting NFRs (I1) i.e. the situation in which developers and/or stakeholders focus on the system's functionality and do not identify NFRs in a sufficient manner, often postponing such task to a later stage of the project. Unfortunately, it often results in significant rework effort, as NFRs are not necessarily simple additions and are likely to substantially affect the system architecture. It should be stated that this issue, while mentioned by many papers, is not always based on experience or empirical findings

Table 1. Problems and challenges affecting development of NFRs in ASD

ID	Problem/challenge	RE	Sources
I1	Neglecting NFRs (usually while focusing on FR)	+	[25], [52], [53], [54], [55], [56], [57], [58], [59], [60], [61], [62], [63], [64], [65], [66], [67]
I2	Misunderstandings regarding NFRs specified as User Stories (or similar simplified representation)	+	[62], [68], [69], [70], [71], [72], [73]
I3	Lack of recognition of NFRs by stakeholders	+	[25], [52], [55], [60], [74], [75]
I4	Difficulties with documenting the NFRs in a way that exposes their dependencies	+	[52], [55], [63], [65], [67]
I5	Lack of traceability mechanisms of NFRs	+	[58], [67], [76], [77]
I6	Inadequate NFR test specification to verify their implementation	+	[52], [57], [74], [78]
I7	Insufficient knowledge/competencies (advanced NFR concepts) in the project team	+	[52], [55], [75]
I8	Overlooking sources of NFRs (stakeholders)	+	[52], [55], [74]
I9	Unclear conceptual definition of NFRs (how to document them)	+	[25], [52], [79]
I10	NFRs are affected by changes in FRs	+	[67], [80], [81]
I11	Sporadic adherence to quality guidelines by Agile teams		[52], [55], [74],
I12	Suboptimal inter-team organization (around components, scenarios or functional teams e.g. testers) leading to poor implementation of NFRs		[52], [55], [74]
I13	Late detection of NFRs' infeasibility	+	[52], [74]
I14	Ambiguous NFRs communication process	+	[52], [67]
I15	NFRs stored outside of backlog, in an external document and thus not always addressed	+	[39], [71]
I16	Hidden assumptions regarding NFRs implementation in inter-team collaboration (in a large scale project)		[52], [74]
I17	Misunderstanding the architecture drivers (priorities of NFRs) between teams		[52], [74]
I18	Lengthy NFR acceptance checklist (e.g. DoD)	+	[52]
I19	Agile process does not include a feedback loop regarding NFRs	+	[55]
I20	Unmanaged architecture changes		[52]
I21	Lack of cost-effective real integration test		[52]
I22	Adopting legacy architectural decisions complicate the implementation of NFRs of the new system		[74]
I23	Moving to Agile with a waterfall mind-set		[74]
I24	Difficult testing to verify NFRs as it requires associated FR to be already implemented		[82]

Table 2. Practices facilitating implementing NFRs in ASD

ID	Practice	RE	Sources
P1	Use modified or additional specification techniques for NFRs (including those adopted from plan-driven approaches)	+	[60], [61], [62], [63], [65], [66], [67], [70], [71], [75]
P2	Maintain traceability between FRs and NFRs	+	[58], [63], [65], [76], [77], [80], [81], [83]
P3	Start focusing on NFRs early in the project	+	[61], [64], [66], [79], [84], [85], [86]
P4	Document NFRs using standard ARE specification techniques (e.g. US, DoD, AC)	+	[55], [56], [71], [73], [84], [87], [88]
P5	Use automated monitoring tools, e.g. SONAR, to monitor quality of software under development		[52], [53], [54], [55], [74], [87], [89]
P6	Involve NFR specialists (e.g. a team of specialists that ensures proper implementation of NFRs or an NFR stakeholder)	+	[52], [55], [57], [67], [74]
P7	Involve multiple roles and viewpoints to elicit and/or review NFRs	+	[62], [66], [67], [78], [90]
P8	Educate and raise awareness about the importance of (particular) NFRs	+	[55], [79], [90]
P9	Use patterns/templates catalogue to specify NFRs	+	[53], [62], [91]
P10	Establish preparation team (responsible for NFRs, architecture and distribution of backlog items to development teams)		[52], [74], [87]
P11	Use abstract but easy to grasp terms by user and/or alternatives to elicit NFRs from stakeholders	+	[78], [84]
P12	Use multiple product backlogs to include requirements of different viewpoints	+	[52], [74]
P13	Use supporting systems providing NFR recommendations	+	[54], [92]
P14	Instead of specifying NFRs as epics, user stories etc., use a similar but distinct structure dedicated to NFRs	+	[69], [93]
P15	Reserve part of the sprint for important NFRs		[52], [74]
P16	Introduce Sprint allocation based on multiple Product Backlogs (e.g. 1 - FRs, 2 - NFRs, 3 - CI/CD requirements)		[52], [74]
P17	Establish components teams (each team solely responsible for a given component and its quality)		[52], [74]
P18	Introduce innovation and planning iteration (IP, term from SAFe) to resolve technical debts related to NFRs		[52], [74]
P19	Conduct NFR-oriented code reviews		[55], [87]
P20	Explain to the stakeholders the consequences of overspecified NFRs	+	[84]
P21	Maintain an assumption wiki-page	+	[52]
P22	Use CI environment to utilize automated NFR testing		[82]
P23	Establish an independent team to test NFRs' implementation		[79]

but sometimes treated as "common knowledge" or quoted from other referenced papers. On the other hand, the frequent occurrence of such issue is confirmed by more general studies not dedicated to NFRs but listing the general problems and challenges related to ASD and/or requirements engineering (examples are given in Sect. 5.1).

A number of issues can be attributed to limitations of the simplified requirements documentation techniques (e.g. user stories, story cards) commonly used in Agile methods. In application to NFRs, such techniques can turn out to be insufficient to express NFRs in an unambiguous way (I2). Another reported shortcoming of such techniques is the difficulty in representing the dependencies between a given NFR and other related requirements (I4). An open issue of how to represent NFRs is also reported as a doubt explicitly expressed by Agile teams (I9). While in some projects, a workaround in the form of a separate document dedicated to NFRs is used, it can also cause difficulties as the project team can focus on the FRs typically stored in the product backlog and do not sufficiently rely on external documents including that for NFRs (I15).

NFRs are more difficult to capture and cause problems both for stakeholders and for the project team. The stakeholders may even not recognize their needs that have to be captured as NFRs (I3). The project team may in turn lack the knowledge and competencies necessary to identify and implement some NFRs, especially when advanced concepts related to e.g. security or performance need to be used (I7).

The elicitation and communication of NFRs is another category of issues. Requirements elicitation can fail to involve all of the relevant stakeholders (I8) and result in NFRs that do not reflect all viewpoints or even omit some important requirements. NFRs are also quite hard to express, thus their communication (both from the stakeholder to the project team, and between team members) can be prone to errors (I14). Moreover, in large scale development projects, involving multiple teams, additional communication problems are likely to arise (I12, I16, I17). Several drawbacks in handling NFRs can result in a situation of late detection of NFR infeasibility (I13), especially considering the lack of a feedback loop regarding NFRs (I19).

Several issues related to NFR traceability and verifiability are reported as well. A lack of NFR traceability mechanisms is claimed in general (I5), but also several more specific issues are described. Traceability of NFRs is even more important as NFRs are frequently affected by changes in FRs (I10). It is difficult to develop test specifications associated with NFRs, which are intended to verify their implementation (I6). Moreover the execution of such tests requires the associated FRs to be already implemented (I24). The cost-effectiveness of some tests is also disputed (I21). The manual verification of DoD can be cumbersome as well, especially in case of a lengthy checklist (I18).

The remaining issues are either related to project team members' attitudes (I11, I23) or architectural design activities (I20, I22).

4.2 What Practices Facilitate the Successful Identification and Implementation of Non-functional Requirements in ASD?

A number of practices dedicated to the documentation of NFRs can be found in the literature, even though some of them seem to be mutually contradictory. The issue of the insufficiency of the popular Agile requirements documentation techniques can be addressed by utilizing modified or additional specification techniques (P1). Such techniques are to be applied to NFRs only (while FRs are still recorded as e.g. user stories). Some proposals include techniques adopted from plan-driven approaches.

Alternatively, other sources recommend making sure that NFRs are documented together with FRs, using the same, typical representations, e.g. user stories, Definition of Done, Acceptance Criteria (P4). There is also a kind of intermediate solution suggested - instead of specifying NFRs as epics, user stories etc. and mixing them with FRs, a similar but distinct structure dedicated to NFRs can be used (P14). Also, assumptions related to the implementation of NFRs are worth documenting using, e.g. a wiki-page (P21).

Focusing on NFRs in an early phase of the project (P3) is a suggestion that can possibly minimize the rework caused by omitted NFRs. Multiple roles and viewpoints should be involved to elicit and/or review NFRs (P7). A number of more detailed practices facilitating requirements elicitation from stakeholders can also be found, e.g. using proper terms (P11) or explaining the consequences of NFRs expressed by stakeholders (P20), especially in the case of "over-specification". Another good idea is to educate and raise awareness about importance of NFRs (P8) - in general or with respect to some categories of NFRs which are not sufficiently recognized. Both stakeholders and project team members can be educated in such a manner.

As NFRs can be difficult to identify and even harder to implement, it is possible to strengthen the competencies of the project team by involving NFR specialists (P6). For example, a security expert can enable the elicitation of relevant security requirements and later verify their implementation. As the stakeholders may not be able to identify the NFRs themselves, external resources such as catalogues of NFR patterns/templates (P9) or dedicated supporting systems providing recommendations (P13) can be used as additional sources of NFRs.

Maintenance of traceability between FRs and NFRs is a frequently recommended practice (P2) that enables proper requirements management activities, including configuration and change management. Various solutions, including tool support, are proposed to ensure traceability maintenance. There are also other practices including the use of automated tools, in particular: tools to monitor the quality of the software under development, including the aspects expressed in NFRs (P5) and CI environments facilitating automated NFR testing (P22). Apart from tool-based testing, NFR-oriented code reviews (P19) and external tests conducted by an independent team (P23) can be practiced.

To minimize the risk of neglecting NFRs, several actions in the software project organization can be undertaken. They can concern the organization of the project team(s) (P10, P17); the development process - dedicating an iteration

(P18) or a part of each iteration (P15) to the implementation of NFRs; or using multiple requirements registers, which make NFRs (or specific NFR categories) more visible (P12, P16).

5   Discussion

Non-functional requirements (NFRs) have become an important research area, mainly due to the abundance of project failures caused by neglecting quality attributes related to user values. While there is no consensus on the reasons for this, Maxim and Kessentini point out that NFRs "are not easy for stakeholders to articulate, but they know that the software will not be usable without some of these non-functional characteristics" [94]. Similarly, the four most frequent issues identified in our study concern neglecting (I1), a lack of (I3), or misunderstanding (I2) NFRs. Further to this, even when one manages to write them down, difficulties are encountered along the way while attempting to document particular qualities and their dependencies (I4).

Notwithstanding these observations, one question arises: Why have been NFRs disregarded? The first reason is the insufficient knowledge and low competence of the employees (I7), particularly in terms of their analytical skills and professional experience, reflected by their inability to perform a required task at a targeted level of proficiency. Moreover, their incompetence is also demonstrated by overlooking sources of NFRs (I8), vague definitions and obscure descriptions (I9). These burdens might be considered to be a result of ambiguous communication (I14).

The remaining aspects of the discussion, namely: comparison to works by other authors (5.1), study limitations (5.2) and implications (5.3) are respectively given below.

5.1   Comparison with Related Works

There are only a few sources dedicated to identification of NFR-related issues and/or practices in ASD. Alsaquaf et al. consider NFRs in a more specific context of Agile Large-Scale Distributed projects. These authors conducted an SLR study to summarize the challenges and practices mentioned in the literature [30] and further investigated additional challenges through a series of interviews with industry practitioners [52]. Behutiye et al. [85] consider NFRs in the generic context of ASD. They used situational method engineering to analyze NFR management practices and interviews to identify challenges and practices of NFR documentation [71].

The SLR by Alsaquaf et al. [30] uncovered 12 NFR-related issues and 13 practices used as solutions. We were able to identify more items of both categories. The results of both above-mentioned primary studies ([52] and [85]) were retrieved in our SLR study and included in its results, together with a wider set of issues/practices from other sources. It is worth to notice that 16 out of 24 issues found in our SLR were identified in a series of interviews dedicated to NFRs challenges in Agile Large-Scale Distributed (ALSD) development [52], which indicates that such challenges are not limited to ALSD, but apply to ASD projects in general. Such findings are also corroborated by reports in other sources we were able to retrieve in our SLR study.

A larger number of research studies on issues and/or practices is available, however their scope is wider and concerns e.g. the challenges of Agile requirements engineering in general. An SLR study by Inayat *et al.* summarizes existing requirements engineering challenges, Agile practices that address such challenges as well as additional challenges related to Agile practices [28]. Heikilla *et al.* identified, through a mapping study, Agile requirements engineering benefits as well as its problematic areas [95]. Medeiros *et al.* conducted a mapping study focused on Agile requirements engineering practices and techniques dedicated to requirements elicitation and documentation [51]. Schon *et al.* focuses on introducing the user's perspective to ASD and joint application of ASD and User-Centered Design (UCD). They conducted an SLR study to summarize the related practices and identify essential aspects of Agile requirements engineering in the UCD context [27].

Apart from [30], [71] and [85], none these studies considers NFRs specifically and as such, they do not address detailed practices nor issues. Some of them enumerate single NFR-related issues, mostly mentioning the risk of neglecting NFRs in ASD projects [27,28] and to the typical Agile documentation insufficient to successfully capture NFRs [28,95]. It is worth noting here that usually, the issues/practices mentioned in such papers are not explicitly assigned to NFRs but to requirements in general, also including FRs, constraints, business goals etc. With respect to this, our study has a much more narrow scope, but within this scope provides much more detailed findings.

Finally, there are studies that focus on a single issue and/or practice, as they propose a new method, technique or practice to address some known issue, e.g. a lack of NFR traceability or difficulty in documenting NFRs of some kind. In our study, we made an attempt to include such contributions into the summary lists that provide the reader with an overview of the state of the art on NFRs in the ASD topic.

5.2   Limitations

While we try to minimize risks by following the guidelines by Kitchenham and Charters [47], we are aware that our study has several possible limitations remaining. A limitation of any SLR study is the potential bias in selecting the sources. This starts with the decisions regarding the publication databases to be searched and the search string to be used for this purpose.

Our study conducted the search in Elsevier Scopus only. Scopus is known to enable a single search query to access items from a large number of journals and conferences [49,50]. However, it is possible that it misses some sources that could in turn be found in other databases. Moreover, scientific databases do not include so called "grey literature" which can potentially include industry experiences in non-scientific publications, such as reports and web articles.

We paid attention to the construction of the search string and included several synonyms and alternative terms to increase our chances of finding all of the relevant sources. The SLR is, however, strongly dependent on vocabulary and we cannot rule out that some authors used less common expressions which would lead to failure to find their papers. Moreover, as our search string implied that NFRs must be explicitly mentioned in the paper's title, abstract or keywords, more generic papers (e.g. dedicated to ASD challenges or practices), which just mention an issue or practice related to NFRs among many others, could be missed.

The extraction of the data from the sources is also a task prone to bias, as it is done by humans, who interpret the contents of the sources. Following SLR guidelines minimizes such a threat, but cannot entirely eliminate it.

5.3 Implications for Research and Practice

This study has several implications for both researchers and practitioners. For researchers, the relatively small number of sources retrieved for the SLR indicates that there is a need for more studies regarding NFRs in ASD projects. Moreover, the issues and practices listed in the findings of our study can be considered by researchers as potential subjects of dedicated empirical studies, further exploring, e.g. the root causes of reported issues or the effectiveness of the facilitating practices.

Industrial practitioners can use our findings to anticipate issues in projects they participate in, and to select facilitating practices to be applied in their projects. Our study can also be used to raise awareness on NFRs in ASD, the related issues and practices as well as the overall importance of such a topic - which still tends to be neglected or lack sufficient attention.

6 Conclusions

In this paper, we explored the topic of the implementation of non-functional requirements (NFRs) in Agile Software Development (ASD), focusing on the issues and facilitating practices, gathered from the existing body of literature. The main motivation underpinning this study was to investigate the state-of-the-art in implementing NFRs within ASD projects, through a systematic literature review (SLR), in order to identify what issues have been documented, as well as by what means one can facilitate the implementation of NFRs. This research was driven by the guidelines elaborated by Kitchenham and Charters, a well-known and widely-applied research framework in the field of software engineering.

The number of industrial projects that deliver different and specific lessons regarding NFRs, makes comparison of the studies a time-consuming and intricate task, since they do not frequently deal with the same focus or goals. Nevertheless, we were able to collect and unambiguously classify a bulk of issues and practices, extracted from peer-reviewed scientific sources. Obviously, our report neither exhausts the topic nor provides external validity. However, we believe that the obtained results may serve as a useful reference repository to be used by both experienced and novice researchers, as well as senior and junior practitioners.

Moreover, a number of open issues and related research directions were identified through this study, which can be considered as an input for future work. A clear lack of consensus on which requirements documentation techniques should be used in order to specify NFRs. Some sources suggest using the techniques available in Agile methods e.g. User Stories (possibly with some adjustments), while others recommend introducing additional techniques (including those used in more traditional, plan-driven approaches). Such selection of the suitable techniques may be a context-dependent issue and requires further investigation. Other areas we identify as potential future work directions are: the relationship between requirements engineering and testing areas (test specifications to verify the implementation of NFRs, which can also be considered as part of NFRs documentation); and the facilitation of an unambiguous NFRs communication process.

We plan to address the latter topic by designing and testing an ontology-based approach, using Controlled Natural Language (CNL) [96] and the Fluent Editor [97], simulating and modelling requirements specification. Our observations gathered during professional work indicate that there is still a need to implement suitable methods and tools to support communication among different groups of stakeholders and development teams.

References

1. Amjad, S., et al.: Calculating completeness of agile scope in scaled agile development. IEEE Access **6**, 5822–5847 (2017)
2. Adnan, M., Afzal, M.: Ontology based multiagent effort estimation system for Scrum agile method. IEEE Access **5**, 25993–26005 (2017)
3. Strandberg, P.E., Enoiu, E.P., Afzal, W., Sundmark, D., Feldt, R.: Information flow in software testing–an interview study with embedded software engineering practitioners. IEEE Access **7**, 46434–46453 (2019)
4. Tjørnehøj, G., Fransgård, M., Skalkam, S.: Trust in agile teams in distributed software development. In: Information System Research Seminar in Scandinavia 2012 Information Systems Research Seminar in Scandinavia, pp. 1–15 (2012)
5. Martin, R.C.: Agile Software Development: Principles, Patterns and Practices. Prentice Hall, Upper Saddle River (2002)
6. Roy, S., Raju, A., Mandal, S.: An empirical investigation on e-retailer agility, customer satisfaction, commitment and loyalty. Bus. Theory Pract. **18**, 97–108 (2017)
7. Consultancy.eu: Half of companies applying agile methodologies & practices (2020). Accessed 10 Nov 2020. https://www.consultancy.eu/news/4153/half-of-companies-applying-agile-methodologies-practices
8. Version One: 13th annual state of agile report (2019). Accessed 10 Nov 2020. https://stateofagile.com/
9. Bjarnason, E., Wnuk, K., Regnell, B.: A case study on benefits and side-effects of agile practices in large-scale requirements engineering. In: 1st Workshop on Agile Requirements Engineering, pp. 1–5 (2011)

10. Kaur, K., Jajoo, A., et al.: Applying agile methodologies in industry projects: benefits and challenges. In: 2015 International Conference on Computing Communication Control and Automation, pp. 832–836. IEEE (2015)
11. Diebold, P., Mayer, U.: On the usage and benefits of agile methods & practices. In: Baumeister, H., Lichter, H., Riebisch, M. (eds.) XP 2017. LNBIP, vol. 283, pp. 243–250. Springer, Cham (2017). https://doi.org/10.1007/978-3-319-57633-6_16
12. Guzmán, L., Oriol, M., Rodríguez, P., Franch, X., Jedlitschka, A., Oivo, M.: How can quality awareness support rapid software development? – a research preview. In: Grünbacher, P., Perini, A. (eds.) REFSQ 2017. LNCS, vol. 10153, pp. 167–173. Springer, Cham (2017). https://doi.org/10.1007/978-3-319-54045-0_12
13. Guamán, D.S., Del Alamo, J.M., Caiza, J.C.: A systematic mapping study on software quality control techniques for assessing privacy in information systems. IEEE Access 8, 74808–74833 (2020)
14. Jarzębowicz, A., Połocka, K.: Selecting requirements documentation techniques for software projects: a survey study. In: 2017 Federated Conference on Computer Science and Information Systems (FedCSIS), pp. 1189–1198. IEEE (2017)
15. Ryan, A.J.: An approach to quantitative non-functional requirements in software development. In: 34th Annual Government Electronics and Information Association Conference, pp. 13–20 (2000)
16. Kautz, K.: Customer and user involvement in agile software development. In: Abrahamsson, P., Marchesi, M., Maurer, F. (eds.) XP 2009. LNBIP, vol. 31, pp. 168–173. Springer, Heidelberg (2009). https://doi.org/10.1007/978-3-642-01853-4_22
17. Jarzębowicz, A., Marciniak, P.: A survey on identifying and addressing business analysis problems. Found. Comput. Decis. Sci. 42(4), 315–337 (2017)
18. Jarzębowicz, A., Ślesiński, W.: What Is troubling IT analysts? A survey report from Poland on requirements-related problems. In: Kosiuczenko, P., Zieliński, Z. (eds.) KKIO 2018. AISC, vol. 830, pp. 3–19. Springer, Cham (2019). https://doi.org/10.1007/978-3-319-99617-2_1
19. Mohammadi, S., Nikkhahan, B., Sohrabi, S.: Challenges of user involvement in extreme programming projects. Int. J. Softw. Eng. Appl. 3(1), 19–32 (2009)
20. Bano, M., Zowghi, D.: A systematic review on the relationship between user involvement and system success. Inf. Softw. Technol. 58, 148–169 (2015)
21. Schmitz, K., Mahapatra, R., Nerur, S.: User engagement in the era of hybrid agile methodology. IEEE Softw. 36(4), 32–40 (2018)
22. Beck, K., et al.: The agile manifesto (2001). Accessed 10 Nov 2020, https://agilemanifesto.org/
23. Leffingwell, D.: Agile Software Requirements: Lean Requirements Practices for Teams, Programs, and the Enterprise. Addison-Wesley Professional, Boston (2010)
24. Miler, J., Gaida, P.: On the agile mindset of an effective team-an industrial opinion survey. In: 2019 Federated Conference on Computer Science and Information Systems (FedCSIS), pp. 841–849. IEEE (2019)
25. Ramesh, B., Cao, L., Baskerville, R.: Agile requirements engineering practices and challenges: an empirical study. Inf. Syst. J. 20(5), 449–480 (2010)
26. Zakrzewski, M., Kotecka, D., Ng, Y.Y., Przybyłek, A.: Adopting collaborative games into agile software development. In: Damiani, E., Spanoudakis, G., Maciaszek, L.A. (eds.) ENASE 2018. CCIS, vol. 1023, pp. 119–136. Springer, Cham (2019). https://doi.org/10.1007/978-3-030-22559-9_6
27. Schön, E.-M., Winter, D., Escalona, M.J., Thomaschewski, J.: Key challenges in agile requirements engineering. In: Baumeister, H., Lichter, H., Riebisch, M. (eds.) XP 2017. LNBIP, vol. 283, pp. 37–51. Springer, Cham (2017). https://doi.org/10.1007/978-3-319-57633-6_3

28. Inayat, I., Salim, S.S., Marczak, S., Daneva, M., Shamshirband, S.: A systematic literature review on agile requirements engineering practices and challenges. Comput. Human Behav. **51**, 915–929 (2015)
29. Soares, H.F., Alves, N.S., Mendes, T.S., Mendonça, M., Spínola, R.O.: Investigating the link between user stories and documentation debt on software projects. In: 2015 12th International Conference on Information Technology-New Generations, pp. 385–390. IEEE (2015)
30. Alsaqaf, W., Daneva, M., Wieringa, R.: Quality requirements in large-scale distributed agile projects – a systematic literature review. In: Grünbacher, P., Perini, A. (eds.) REFSQ 2017. LNCS, vol. 10153, pp. 219–234. Springer, Cham (2017). https://doi.org/10.1007/978-3-319-54045-0_17
31. Svensson, R.B., Gorschek, T., Regnell, B., Torkar, R., Shahrokni, A., Feldt, R.: Quality requirements in industrial practice–an extended interview study at eleven companies. IEEE Trans. Softw. Eng **38**(4), 923–935 (2011)
32. Umar, M., Khan, N.A.: Analyzing non-functional requirements (NFRs) for software development. In: 2011 IEEE 2nd International Conference on Software Engineering and Service Science, pp. 675–678. IEEE (2011)
33. Zhang, X., Wang, X.: Tradeoff analysis for conflicting software non-functional requirements. IEEE Access **7**, 156463–156475 (2019)
34. Weichbroth, P.: Delivering usability in IT products: empirical lessons from the field. Int. J. Softw. Eng. Knowl. Eng. **28**(07), 1027–1045 (2018)
35. Suryawanshi, T., Rao, G.: A survey to support NFRs in agile software development process. Int. J. Comput. Sci. Inf. Technol. **6**(6), 5487–5489 (2015)
36. Rosa, N.S., Justo, G.R., Cunha, P.R.: A framework for building non-functional software architectures. In: 2001 ACM Symposium on Applied Computing, pp. 141–147 (2001)
37. Mizouni, R., Salah, A.: Towards a framework for estimating system NFRs on behavioral models. Knowl.-Based Syst. **23**(7), 721–731 (2010)
38. Charette, R.N.: The biggest IT failures of 2018 (2018). Accessed 18 Sept 2020. https://spectrum.ieee.org/riskfactor/computing/it/it-failures-2018-all-the-old-familiar-faces
39. Maiti, R.R., Mitropoulos, F.J.: Capturing, eliciting, predicting and prioritizing (CEPP) non-functional requirements metadata during the early stages of agile software development. In: SoutheastCon 2015, pp. 1–8. IEEE (2015)
40. Statista: Largest software and programming companies worldwide by sales revenue from 2017 to 2020 (2017). Accessed 10 Nov 2020. https://www.statista.com/statistics/790179/worldwide-largest-software-programming-companies-by-sales/
41. Microsoft. Build for the needs of the business (2020). Accessed 19 Sept 2020. https://docs.microsoft.com/en-us/azure/architecture/guide/design-principles/build-for-business
42. Oracle. Best practices for WLI application life cycle (2020). Accessed 10 Nov 2020. https://docs.oracle.com/cd/E13214_01/wli/docs102/bestpract/requirements.html
43. Ossowska, K., Szewc, L., Weichbroth, P., Garnik, I., Sikorski, M.: Exploring an ontological approach for user requirements elicitation in the design of online virtual agents. In: Wrycza, S. (ed.) SIGSAND/PLAIS 2016. LNBIP, vol. 264, pp. 40–55. Springer, Cham (2016). https://doi.org/10.1007/978-3-319-46642-2_3
44. Redlarski, K.: The impact of end-user participation in IT projects on product usability. In: 2013 International Conference on Multimedia, pp. 1–8. Interaction, Design and Innovation (MIDI) (2013)

45. Redlarski, K., Weichbroth, P.: Hard lessons learned: delivering usability in IT projects. In: 2016 Federated Conference on Computer Science and Information Systems (FedCSIS), pp. 1379–1382. IEEE (2016)
46. Buchan, J., Bano, M., Zowghi, D., MacDonell, S., Shinde, A.: Alignment of stakeholder expectations about user involvement in agile software development. In: 21st International Conference on Evaluation and Assessment in Software Engineering, pp. 334–343 (2017)
47. Kitchenham, B., Charters, S.: Guidelines for performing systematic literature reviews in software engineering. Technical Report EBSE-2007-01 (2007)
48. Jarzębowicz, A., Weichbroth, P.: A qualitative study on non-functional requirements in agile software development. Submitted, under review (2020)
49. Scopus: Scopus content coverage guide (2020). Accessed 10 Nov 2020. https://www. elsevier.com/__data/assets/pdf_file/0007/69451/Scopus_ContentCoverage_ Guide_WEB.pdf
50. Daneva, M., Damian, D., Marchetto, A., Pastor, O.: Empirical research methodologies and studies in requirements engineering: how far did we come? J. Syst. Softw. **95**, 1–9 (2014)
51. Medeiros, J., Alves, D.C., Vasconcelos, A., Silva, C., Wanderley, E.: Requirements engineering in agile projects: a systematic mapping based in evidences of industry. In: XVIII Ibero-American Conference on Software Engineering (CIBSE), pp. 460–476 (2015)
52. Alsaqaf, W., Daneva, M., Wieringa, R.: Quality requirements challenges in the context of large-scale distributed agile: an empirical study. Inf. Softw. Technol. **110**, 39–55 (2019)
53. Oriol, M., et al.: Data-driven elicitation of quality requirements in agile companies. In: Piattini, M., Rupino da Cunha, P., García Rodríguez de Guzmán, I., Pérez-Castillo, R. (eds.) QUATIC 2019. CCIS, vol. 1010, pp. 49–63. Springer, Cham (2019). https://doi.org/10.1007/978-3-030-29238-6_4
54. Ramos, F.B.A., Costa, A.A.M., Perkusich, M., Almeida, H.O., Perkusich, A.: A non-functional requirements recommendation system for Scrum-based projects. In: 30th International Conference on Software Engineering & Knowledge Engineering (SEKE), pp. 149–148 (2018)
55. Terpstra, E., Daneva, M., Wang, C.: Agile practitioners'understanding of security requirements: insights from a grounded theory analysis. In: 25th International Requirements Engineering Conference Workshops (REW), pp. 439–442. IEEE (2017)
56. Sachdeva, V., Chung, L.: Handling non-functional requirements for big data and IOT projects in Scrum. In: 7th International Conference on Cloud Computing, Data Science & Engineering-Confluence, pp. 216–221. IEEE (2017)
57. Schön, E.-M., Thomaschewski, J., Escalona, M.J.: Agile requirements engineering: a systematic literature review. Comput. Stand. Interfaces **49**, 79–91 (2017)
58. Aljallabi, B.M., Mansour, A.: Enhancement approach for non-functional requirements analysis in agile environment. In: 2015 International Conference on Computing, Control, Networking, Electronics and Embedded Systems Engineering (ICC-NEEE), pp. 428–433. IEEE (2015)
59. Käpyaho, M., Kauppinen, M.: Agile requirements engineering with prototyping: a case study. In: 23rd International Requirements Engineering Conference (RE), pp. 334–343. IEEE (2015)
60. Domah, D., Mitropoulos, F.J.: The NERV methodology: a lightweight process for addressing non-functional requirements in agile software development. In: SoutheastCon 2015, pp. 1–7. IEEE (2015)

61. Dragicevic, S., Celar, S., Novak, L.: Use of method for elicitation, documentation, and validation of software user requirements (MEDoV) in agile software development projects. In: 6th International Conference on Computational Intelligence, Communication Systems and Networks, pp. 65–70. IEEE (2014)

62. Nawrocki, J., Ochodek, M., Jurkiewicz, J., Kopczyńska, S., Alchimowicz, B.: Agile requirements engineering: a research perspective. In: Geffert, V., Preneel, B., Rovan, B., Štuller, J., Tjoa, A.M. (eds.) SOFSEM 2014. LNCS, vol. 8327, pp. 40–51. Springer, Cham (2014). https://doi.org/10.1007/978-3-319-04298-5_5

63. Farid, W.M., Mitropoulos, F.J.: Visualization and scheduling of non-functional requirements for agile processes. In: SoutheastCon 2013, pp. 1–8. IEEE (2013)

64. Bourimi, M., Kesdogan, D.: Experiences by using AFFINE for building collaborative applications for online communities. In: Ozok, A.A., Zaphiris, P. (eds.) OCSC 2013. LNCS, vol. 8029, pp. 345–354. Springer, Heidelberg (2013). https://doi.org/10.1007/978-3-642-39371-6_39

65. Farid, W., Mitropoulos, F.: Novel lightweight engineering artifacts for modeling non-functional requirements in agile processes. In: SoutheastCon 2012, pp. 1–7. IEEE (2012)

66. Um, T., Kim, N., Lee, D., In, H.P.: A quality attributes evaluation method for an agile approach. In: 1st ACIS/JNU International Conference on Computers, Networks, Systems and Industrial Engineering, pp. 460–461. IEEE (2011)

67. Bourimi, M., Barth, T., Haake, J.M., Ueberschär, B., Kesdogan, D.: AFFINE for enforcing earlier consideration of NFRs and human factors when building sociotechnical systems following agile methodologies. In: Bernhaupt, R., Forbrig, P., Gulliksen, J., Lárusdóttir, M. (eds.) HCSE 2010. LNCS, vol. 6409, pp. 182–189. Springer, Heidelberg (2010). https://doi.org/10.1007/978-3-642-16488-0_15

68. Boehm, B., Rosenberg, D., Siegel, N.: Critical quality factors for rapid, scalable, agile development. In: 19th International Conference on Software Quality, Reliability and Security Companion (QRS-C), pp. 514–515. IEEE (2019)

69. Ionita, D., van der Velden, C., Ikkink, H.J.K., Neven, E., Daneva, M., Kuipers, M.: Towards risk-driven security requirements management in agile software development. In: Cappiello, C., Ruiz, M. (eds.) Information Systems Engineering in Responsible Information Systems, CAiSE 2019. Lecture Notes in Business Information Processing, vol. 350, pp. 133–144. Springer, Heidelberg (2019). https://doi.org/10.1007/978-3-030-21297-1_12

70. Medeiros, J., Vasconcelos, A., Goulão, M., Silva, C., Araújo, J.: An approach based on design practices to specify requirements in agile projects. In: ACM Symposium on Applied Computing, pp. 1114–1121 (2017)

71. Behutiye, W., Karhapää, P., Costal, D., Oivo, M., Franch, X.: Non-functional requirements documentation in agile software development: challenges and solution proposal. In: Felderer, M., Méndez Fernández, D., Turhan, B., Kalinowski, M., Sarro, F., Winkler, D. (eds.) PROFES 2017. LNCS, vol. 10611, pp. 515–522. Springer, Cham (2017). https://doi.org/10.1007/978-3-319-69926-4_41

72. Patel, C., Ramachandran, M.: Story Card Maturity Model (SMM): a process improvement framework for agile requirements engineering practices. J. Softw. (JSW) 4(5), 422–435 (2009)

73. Patel, C., Ramachandran, M.: Bridging best traditional SWD practices with XP to improve the quality of XP projects. In: International Symposium on Computer Science and its Applications, pp. 357–360. IEEE (2008)

74. Alsaqaf, W., Daneva, M., Wieringa, R.: Understanding challenging situations in agile quality requirements engineering and their solution strategies: insights from a case study. In: 26th International Requirements Engineering Conference (RE), pp. 274–285. IEEE (2018)
75. Younas, M., Jawawi, D., Ghani, I., Kazmi, R.: Non-functional requirements elicitation guideline for agile methods. J. Telecommun. Electron. Comput. Eng. (JTEC) 9(3–4), 137–142 (2017)
76. Jawawi, D., Arbain, A., Kadir, W., Ghani, I.: Requirement traceability model for agile development: results from empirical studies. Int. J. Innov. Technol. Explor. Eng. 8(8S), 402–405 (2019)
77. Arbain, A.F., Jawawi, D.N.A., Ghani, I., Kadir, W.M.W.: Non-functional requirement traceability process model for agile software development, J. Telecommun. Electron. Comput. Eng. (JTEC) 9(3–5), 203–211 (2017)
78. Macasaet, R.J., Chung, L., Garrido, J.L., Noguera, M., Rodríguez, M.L.: An agile requirements elicitation approach based on NFRs and business process models for micro-businesses. In: 12th International Conference on Product-Focused Software Development and Process Improvement (PROFES), pp. 50–56 (2011)
79. Ambler, S.W.: Beyond functional requirements on agile projects-strategies for addressing nonfunctional requirements. Dr. Dobb's J. (2008)
80. Firdaus, A., Ghani, I., Jawawi, D.N.A., Kadir, W.M.N.W.: Non functional requirements (NFRs) traceability metamodel for agile development. Jurnal Teknologi 77(9) (2015)
81. Arbain, A.F.B., Ghani, I., Kadir, W.M.N.W.: Agile non functional requirements (NFR) traceability metamodel. In: 8th Malaysian Software Engineering Conference (MySEC), pp. 228–233. IEEE (2014)
82. Yu, L., Alégroth, E., Chatzipetrou, P., Gorschek, T.: Utilising CI environment for efficient and effective testing of NFRs. Inf. Softw. Technol. 117, 106199 (2020)
83. Sinnhofer, A.D., Oppermann, F.J., Potzmader, K., Orthacker, C., Steger, C., Kreiner, C.: Increasing the visibility of requirements based on combined variability management. In: Shishkov, B. (ed.) BMSD 2018. LNBIP, vol. 319, pp. 203–220. Springer, Cham (2018). https://doi.org/10.1007/978-3-319-94214-8_13
84. Kopczyńska, S., Ochodek, M., Nawrocki, J.: On importance of non-functional requirements in agile software projects—a survey. In: Jarzabek, S., Poniszewska-Marańda, A., Madeyski, L. (eds.) Integrating Research and Practice in Software Engineering. SCI, vol. 851, pp. 145–158. Springer, Cham (2020). https://doi.org/10.1007/978-3-030-26574-8_11
85. López, L., Behutiye, W., Karhapää, P., Ralyté, J., Franch, X., Oivo, M.: Agile quality requirements management best practices portfolio: a situational method engineering approach. In: Felderer, M., Méndez Fernández, D., Turhan, B., Kalinowski, M., Sarro, F., Winkler, D. (eds.) PROFES 2017. LNCS, vol. 10611, pp. 548–555. Springer, Cham (2017). https://doi.org/10.1007/978-3-319-69926-4_45
86. Alsaqaf, W.: Engineering quality requirements in large scale distributed agile environment. In: International Working Conference on Requirements Engineering: Foundation for Software Quality (REFSQ) Workshops (2016)
87. Mohagheghi, P., Aparicio, M.E.: An industry experience report on managing product quality requirements in a large organization. Inf. Softw. Technol. 88, 96–109 (2017)
88. Silva, A., Araújo, T., Nunes, J., Perkusich, M., Dilorenzo, E., Almeida, H., Perkusich, A.: A systematic review on the use of definition of done on agile software development projects. In: 21st International Conference on Evaluation and Assessment in Software Engineering (EASE), pp. 364–373 (2017)

89. López, L., et al.: Q-rapids tool prototype: supporting decision-makers in managing quality in rapid software development. In: Mendling, J., Mouratidis, H. (eds.) CAiSE 2018. LNBIP, vol. 317, pp. 200–208. Springer, Cham (2018). https://doi.org/10.1007/978-3-319-92901-9_17

90. Camacho, C.R., Marczak, S., Cruzes, D.S.: Agile team members perceptions on non-functional testing: influencing factors from an empirical study. In: 11th International Conference on Availability, Reliability and Security (ARES), pp. 582–589. IEEE (2016)

91. Franch, X., et al.: Data-driven elicitation, assessment and documentation of quality requirements in agile software development. In: Krogstie, J., Reijers, H.A. (eds.) CAiSE 2018. LNCS, vol. 10816, pp. 587–602. Springer, Cham (2018). https://doi.org/10.1007/978-3-319-91563-0_36

92. Ramos, F.B.A., et al.: Evaluating software developers' acceptance of a tool for supporting agile non-functional requirement elicitation. In: 31st International Conference on Software Engineering & Knowledge Engineering (SEKE), pp. 26–42 (2019)

93. Pecchia, C., Trincardi, M., Di Bello, P.: Expressing, managing, and validating user stories: experiences from the market. In: Ciancarini, P., Sillitti, A., Succi, G., Messina, A. (eds.) Proceedings of 4th International Conference in Software Engineering for Defence Applications. AISC, vol. 422, pp. 103–111. Springer, Cham (2016). https://doi.org/10.1007/978-3-319-27896-4_9

94. Maxim, B.R., Kessentini, M.: An introduction to modern software quality assurance. In: Software Quality Assurance, pp. 19–46. Elsevier (2016)

95. Heikkilä, V.T., Damian, D., Lassenius, C., Paasivaara, M.: A mapping study on requirements engineering in agile software development. In: 41st Euromicro SEAA Conference, pp. 199–207. IEEE (2015)

96. Kapłański, P.: Controlled English interface for knowledge bases. Studia Informatica 32(2A), 485–494 (2011)

97. Weichbroth, P.: Fluent editor and controlled natural language in ontology development. Int. J. Artif. Intell. Tools 28(04), 1940007 (2019)

Product Roadmapping Processes for an Uncertain Market Environment: A Grey Literature Review

Stefan Trieflinger[1]([⊠]), Jürgen Münch[1], Jan Schneider[1], Emre Bogazköy[1], Patrick Eißler[1], Bastian Roling[2], and Dominic Lang[3]

[1] Reutlingen University, Alteburgstraße 150, 72762 Reutlingen, Germany
{stefan.trieflinger,juergen.muench}@reutlingen-university.de,
{jan_philip.schneider,emre.bogazkoey,
patrick_denis.eissler}@student.reutlingen-university.de
[2] Viastore Software GmbH, Magirusstraße 13, 70469 Stuttgart, Germany
b.roling@viastore.com
[3] Robert Bosch GmbH, Hoferstraße 30, 71636 Ludwigsburg, Germany
Dominic.lang2@bosch.com

Abstract. Context: Currently, most companies apply approaches for product roadmapping that are based on the assumption that the future is highly predicable. However, nowadays companies are facing the challenge of increasing market dynamics, rapidly evolving technologies, and shifting user expectations. Together with the adaption of lean and agile practices it makes it increasingly difficult to plan and predict upfront which products, services or features will satisfy the needs of the customers. Therefore, they are struggling with their ability to provide product roadmaps that fit into dynamic and uncertain market environments and that can be used together with lean and agile software development practices. **Objective:** To gain a better understanding of modern product roadmapping processes, this paper aims to identify suitable processes for the creation and evolution of product roadmaps in dynamic and uncertain market environments. **Method:** We performed a Grey Literature Review (GLR) according to the guidelines from Garousi et al. **Results:** 32 approaches to product roadmapping were identified. Typical characteristics of these processes are the strong connection between the product roadmap and the product vision, an emphasis on stakeholder alignment, the definition of business and customer goals as part of the roadmapping process, a high degree of flexibility with respect to reaching these goals, and the inclusion of validation activities in the roadmapping process. An overall goal of nearly all approaches is to avoid waste by early reducing development and business risks. From the list of the 32 approaches found, four representative roadmapping processes are described in detail.

Keywords: Product roadmap · Product roadmapping · Product management · Agile methods · Product strategy · UX strategy

© Springer Nature Switzerland AG 2021
A. Przybyłek et al. (Eds.): LASD 2021, LNBIP 408, pp. 111–129, 2021.
https://doi.org/10.1007/978-3-030-67084-9_7

1 Introduction

For the success of a company is it essential to provide a strategic direction in which product offerings will be developed over time to achieve a corporate vision. Product roadmaps can be used as a mechanism to develop and describe such a strategic direction [1]. Cooper and Edge [2] define product roadmaps as a tool that lays out the major initiatives and platforms a business will undertake in the future [3, 4]. The purpose of a product roadmap is to provide an essential understanding, proximity, direction, and some degree of certainty regarding the future direction of the development of a product or a product portfolio [1].

The process to create a product roadmap is called product roadmapping [5]. Figure 1 embeds the product roadmapping process into an overall macro process. Here, the product strategy is derived from the product vision. The product vision mainly describes the reason for creating a product. One level below, a product strategy can be seen as an approach to make or keep a product successful. Consequently, a product strategy should include a product's target group, the key needs to be addressed, the stand-out features of the solution, and the most important business goals. On this basis, the development of the product roadmap can take place. The aim of a product roadmap is to show how the product strategy is put into action. Moreover, it provides the context for making tactical decisions such as deriving the content of the product backlog and its prioritization.

In the following, selected product roadmapping processes are described. It should be noted that a product roadmapping process usually requires customization as every company, product, and set of stakeholders is different [6]. It should be stressed that this article refers to product roadmaps and not to roadmaps in general.

Fig. 1. Product strategy process according to Pichler [6]

A recent study [7] on the state of the practice has shown that the most common product roadmap approach of many software-intensive companies (most of them applying agile and lean practices) consists mainly of specific products, features, or services together with precise release dates for long time horizons (usually one year). This approach can be characterized as feature-based roadmaps, i.e., feature-by-feature wish lists. Feature-based roadmaps are created to inform stakeholders or customers about the point in time a product, feature or service is expected to be delivered [8]. This approach works well in market environments that are predictable, stable, and reliable [7]. However, the market environment often has changed and is now dynamic, complex and uncertain [9]. The main reasons therefore are the high availability of knowledge and resources due to the globalization, rapidly evolving technologies and fast changing customer behaviors [8]. Together with the adoption of lean and agile practices this situation makes it increasingly difficult to plan and predict upfront which products, services, or features should be developed, especially in the mid- and long-term [3, 7]. Thus, companies are struggling more and more with their ability to create reliable product roadmaps [7, 8]. As a result of the mismatch between feature-based roadmaps and dynamic and uncertain market environments, most companies have realized that new approaches and procedures regarding the development and handling of product roadmaps are necessary. Consequently, they have to find new approaches to improve their current product roadmapping capabilities [4].

It should be noted that the Grey Literature Review (GLR) presented in this article is part of a comprehensive GLR on the topic "product roadmapping in a dynamic and uncertain market environment". This means that we searched for relevant articles using a broad search string. After applying our defined inclusion and exclusion criteria, we obtained 170 relevant articles. The analysis of the relevant articles showed that these articles can be divided into the five subcategories: 1) product roadmapping processes, 2) product roadmap alignment, 3) product roadmap formats, 4) product roadmap prioritization techniques, and 5) challenges and pitfalls regarding product roadmapping. We therefore decided to slice the presentation of the results among different articles. We have already published the two papers "Product Roadmap Alignment – Achieving the Vision Together" [10] and the paper "Product Roadmap Formats for an Uncertain Future: A Grey Literature Review" [11]. The latter paper aims to identify structures and contents of a product roadmap that are suitable for operating in a dynamic and uncertain market environment, while the former paper focuses on the identification of measures, methods and techniques that help companies to achieve alignment around the product roadmap. In contrast, the paper at hand strives for identifying suitable processes that provide information on how to create and update product roadmaps in a dynamic and uncertain market environment. Since these papers are outcomes of the broad search and analysis mentioned above, the section "research approach" and the section "threats of validity" are similar to those papers.

The main reason for conducting a grey literature review is the following: a systematic review of the scientific literature on product roadmapping [12] revealed that the available scientific literature offers only little knowledge which processes are suitable to produce a product roadmap for dynamic market environments. To fill this gap and to support companies in improving their product roadmapping processes, this paper aims at identifying such processes based on the analysis of the so-called "grey literature" (i.e., white papers, articles, blogs, business books etc.). The "grey literature" promises to provide more information on how to develop product roadmaps in dynamic market environments. One reason is that companies and experts often communicate their experiences more easily and quickly through grey literature than through scientific literature. Detailed information why we have chosen the grey literature review as research method can be found in Subsect. 3.1 in the section "identification of the need of a GLR".

The outline of this paper is as follows: Sect. 2 presents related work. Section 3 discusses the research approach including a description of the search strategy, the research questions, the search string we used, the applied inclusion and exclusion criteria, the applied selection process as well as the quality assessment we performed. Afterwards, the results of the study are described, and the threats to validity are discussed. Finally, a summary is given.

2 Related Work

Kappel defines roadmaps as forecasts of what is possible or likely to happen, as well as plans that articulate a course of action [13]. In a similar way, DeGregorio [14] points out that roadmaps are visualizations of a forecast, which can be applied in several key areas such as technology, capability, parameter, feature, product, platform, system, environment or threat and business opportunity. Albright [15] considers a roadmap as a document that describes a future environment, objectives to be achieved within that environment, and plans for how those objectives will be achieved over time. Besides, the author points out to review and update a roadmap over time. Otherwise, it is not seen as useful [15]. The verb "roadmapping" describes the process of roadmap development [1]. According to Phaal et al. [16], the most suitable roadmapping process for a company depends on many factors such as the level of available resources (e.g., people, time, budget), the nature of the issue being addressed (e.g., purpose and scope), the available information regarding the market and technology, and other relevant processes and management methods (e.g., strategy, new product development, project management, and market research). Typically, the practice of roadmapping involves social mechanisms, as this process brings together stakeholders from different functions of the organization to plan and make decisions. The roadmap then represents the decisions made [1, 5]. The roadmapping process will differ from one company to the other because companies serve different markets and have different cultures [17].

In the following, examples for conducting the roadmapping process are sketched, which can be found in the scientific literature:

Vähäniitty et al. [18] developed a four-step process for creating and updating product roadmaps, especially in small companies. The process consists of the phases 1) define the strategic mission and vision of the company and outline the product vision, 2) scan the environment, 3) revise and distill the product vision as product roadmaps, and 4) estimate the product life cycle and evaluate the mix of development efforts planned. The author points out that the steps in the process should be performed periodically to adjust the roadmap to new information and changing market situations. Minor updates should also be made to ensure that the roadmap always contains up-to-date information.

Fenwick et al. [19] present an approach to technology roadmapping by integrating marketing and decision-making methodologies. Value drivers are determined to reflect the customer's current needs and future expected needs. These drivers lead to a technology roadmap, which defines technologies to purchase, lease, or develop. The process of developing a value-driven technology roadmap consists of the phases: 1) assessment (evaluation of the company's internal capabilities as well as the external industry environment), 2) market analysis (the understanding of the value proposition for customers), 3) service availability (the creation of an offer of desirable products and services and necessary technologies) and 4) roadmap which is created to link technology to future market opportunities.

Beeton et al. [20] present a roadmapping process to capture and structure insights of supply chains and to develop future views of the competitive issues facing a diverse industrial area. The process includes three main steps, with each main step comprising several sub steps: 1) planning (establish a steering committee, articulate the need for the roadmap, set system boundaries, design the roadmap architecture, recruit experts and miscellaneous preparatory work), 2) insight collection (choose a workshop format, characterize the strategic landscape, conduct a voting process to identify issues from the content of the strategic landscape and rank the identified issues), and 3) insight processing (collate and transcript the insights collected in the roadmapping workshop, develop of visual representations and a working document). The application of the process aims at producing a roadmap that provides useful information, structure and context for strategic planning and innovation in a complex multi-stakeholder industry.

The existing studies that we identified do not explicitly address how to conduct product roadmapping in order to fulfill the requirements of a dynamic and uncertain market environment. There is a lack of documented practical experience and best practices regarding product roadmapping processes that are suitable for the operation in dynamic and uncertain market environments. This is the focus of this study.

An exception is the "design roadmapping process" proposed by Kim et al. [21]. It aggregates design experience elements along a timeline and associates key user needs with the products, services, and/or systems the organization wishes to deliver. The process consists of the steps 1) gather comprehensive user data, experience, and trends, 2) extract core design principles from the user needs, experience and trends, 3) gather an exhaustive list of technologies containing core feature sets of the design concept and prioritize them, 4) map projects to design principles, and 5) create the design roadmap. The author emphasizes that the process should encourage teams to focus on experience-driven planning early, thereby increasing the likelihood of a product desired by the customers. However, the findings of this study are derived from only one case study and the concept of a design roadmap is not directly comparable with the concept of a product roadmap.

3   Research Approach

As this study aims to gain new insights, it was conducted exploratively. To conduct the study in a systematic and repeatable manner, it follows the guidelines according to Garousi et al. [22], which consider three main phases: 1) planning the review, 2) conducting the review and 3) reporting the review (see Table 1).

Table 1. Design of the grey literature review

Planning the review	• Identification of the need of a GLR
	• Formulation of the research questions and scope of the study
	• Definition and refinement of the search string
	• Determination of the inclusion and exclusion criteria
Conducting the review	• Performance of the study selection process
	• Data extraction and conduction of the quality assessment
Reporting the review	• Writing down the findings as documentation (see Sect. 4)

3.1 Planning the Review

Identification of the Need of a GLR: First, we assessed whether a GLR is the appropriate method for our study. Therefore we used the checklist according to Garousi et al. [22] as shown in Table 2. The authors of the checklist propose that if one or more questions can be answered positively, the conduction of a GLR is recommended, otherwise a Systematic Literature Review should be performed. Table 2 shows our answers regarding this study. The first question has been answered by a Systematic Literature Review [12]. This review revealed that the scientific articles describe product roadmapping on a quite abstract level and do not address the demands of an increasingly digital and dynamic market environment with high uncertainty. Based on the checklist, a Grey Literature Review is an appropriate research approach. Furthermore, an initial review of the grey literature and the conduction of expert interviews [7] indicate that there is a high level of interest in insights about the topic "processes for product roadmapping in dynamic and uncertain market environments". Therefore, a Grey Literature Review contributes to the transfer of practical knowledge to the scientific community and practitioners in industry.

Table 2. Checklist according to Garousi et al. [22] to decide whether a grey literature review should be performed

ID	Question	Answer
1	Is the subject "complex" and not solvable by considering only the formal literature?	Yes
2	Is there a lack of volume or quality of evidence, or a lack of consensus of outcome measurement in the formal literature?	Yes
3	Is the contextual information relevant to the subject under study?	Yes
4	Is it the goal to validate or corroborate scientific outcomes with practical experiences?	No
5	Is it the goal to challenge assumptions or falsify results from practice using academic research or vice versa?	No
6	Would a synthesis of insights and evidence from the industrial and academic community be useful to one or even both communities?	Yes
7	Is there a large volume of practitioner sources indicating high practitioner interest in a topic?	Yes

Research Question and Scope of the Study: Our study focuses on identifying suitable processes to create and evolve product roadmaps in dynamic and uncertain market environments. Based on this objective, the following research question was defined:

- RQ1: Which processes are reported in the grey literature to create and evolve a product roadmap in a dynamic and uncertain market environment?

Identification of the Search String: The initial set of our search string was developed in a brainstorming session that aimed at identifying grey literature about product roadmaps in general. The reason is that we wanted to do a broader analysis of the grey literature that does not only focus on the topic "product roadmapping processes". Based on this broader analysis, further analyses with respect to different specific topics were derived. Therefore, "processes" is not part of the search string. To obtain sufficient results and cover our objectives we evolved the search term iteratively. After evaluating different options, we have defined the following search terms:

A1: Innovation, A2: Product*, A3: Product Management, A4: Agile, A5: Outcome* driven, A6: Outcome* oriented, A7: Goal* oriented, A8: Theme*, A9: Roadmap*

The complete search string used in our study was:

(A1 OR A2 OR A3 OR A4 OR A5 OR A6 OR A7 OR A8) AND A9

At the end of the search process we filtered all results that fit to the topic "product roadmapping processes".

Definition of the Inclusion and Exclusion Criteria: In order to filter relevant from irrelevant articles, we defined the following inclusion and exclusion criteria as shown in Table 3.

Table 3. Inclusion and exclusion criteria

Inclusion	• The article discusses the application of product roadmapping in practice
	• The article was published in English
	• The URL is working and freely available
Exclusion	• The source is non text-based
	• The article contains duplicated content of a previously examined article
	• The article is not related to software engineering

3.2 Conducting the Review

Study Selection Process: The data retrieval process was performed by using the predefined search string and applying it to the Google search engine (google.com). To avoid biased results based on past activities the search was conducted in the incognito mode of the browser. Further, a VPN service was used to anonymize the location from which the search was conducted. Moreover, the relevance ranking was applied, which ranks the results according to the Google PageRank algorithm. These steps intend that a minimum of influence of historical search could affect the results. To increase the amount of available URL's the Google option to include similar results was activated. The search was conducted on January 17th, 2020 and yielded in 426 hits. In addition to the search process, we conducted snowballing (i.e., considering further articles that are recommended in an article). This led to 53 further articles. After the application of the selection process we obtained 170 relevant articles which address the main topic product roadmapping.

On this basis we have categorized these 170 articles according to their different subject areas (product roadmapping processes, product roadmap formats, product roadmap prioritization techniques, alignment of different stakeholders around the product roadmap, and challenges and pitfalls regarding product roadmapping).

This led to 32 relevant articles that deal with the topic "product roadmapping processes". The list of these 32 articles can be found in the Appendix. Moreover a detailed overview of what the different approaches focus on can be found on Figshare [23]. Based on an analysis of these 32 articles, four essential approaches that represent typical roadmapping processes are presented in more detail in the results section of this paper. Our applied search process is shown in Fig. 2.

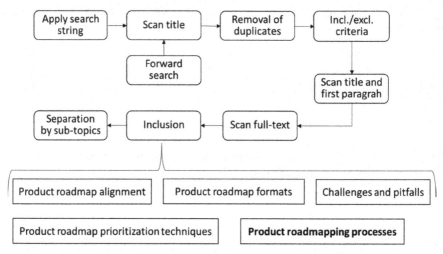

Fig. 2. Study selection process

Quality Assessment and Data Extraction: The criterion for the quality assessment was that the authors were able to comprehend the suggested approach based on their practical experience. The practical experience of each reviewer is shown in Table 4. In the case that the individual reviews led to different results, the process was carried out by a third reviewer to make an inclusion/exclusion decision. Afterwards the results were presented to one practitioner of the Robert Bosch GmbH, who has 7 years of experience regarding product roadmapping and is also co-author of this article. This review did not lead to the exclusion of an identified approach. In the next step, a data extraction was conducted by performing a content analysis for each article and extracting the information needed to answer our research questions. This data extraction serves as input for the reporting, i.e., we documented the findings of each included article (see Sect. 4).

Table 4. Practical experience of the authors regarding product roadmapping

Author	Practical experience regarding product roadmapping
Author 1	8 years
Author 2	3 years
Author 3	2 years
Author 4	1,5 years
Author 5	1 year
Author 6	1 year
Author 7	7 years

4 Results

To answer our research question, we analyzed the selected articles and identified suitable processes that can be used to create and work with a product roadmap in a dynamic and uncertain market environment. A list of the 32 relevant articles can be found in the Appendix. Moreover a detailed overview of what the different approaches focus on can be found on Figshare [23]. Each of these articles deals with an approach for creating and evolving a product roadmap in a dynamic and uncertain market environment. Most of the product roadmapping processes that were found in the grey literature are segmented into three to seven steps. Figure 3 shows the frequency of important components that are described in the identified articles. Most of the product roadmapping practices that were found in the grey literature include the steps 1) setting goals, 2) uncovering and selecting customer needs, 3) assessing the impacts of the feature to be developed to the customer and business goals, and 4) prioritization methods. In contrast the identified articles provide less information about the topics "discovering and experimenting" and "MVP creation and solution building". Moreover the processes found are not sequential but describe the development of a product roadmap and related activities as an iterative process. The product roadmap itself is treated as a variable in these processes, that is, it can change over time.

From this list of the 32 relevant articles, the authors have selected four essential approaches, which fulfill essential requirements for product roadmapping in uncertain and dynamic market environments. The requirements that were used for this selection are derived from an empirical study and described in Trieflinger et al. [24]. In addition, it should be noted that the identified processes are recommendations for the conduction of product roadmapping. Thus, the approaches might need to be adapted to the respective business context or corporate culture.

Lombardo et al. [3] propose a comprehensive process which results in so-called theme-based roadmaps. In order to create a theme-based roadmap the authors recommend using primarily the following components: 1) product vision, 2) business objectives, 3) themes (i.e., high-level customer or system needs) and 4) timeframes. The detailed steps of the process are described in the following:

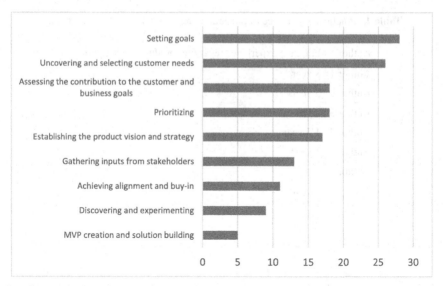

Fig. 3. Frequency of important steps mentioned in the identified articles

1. *Gathering inputs:* The aim of this step is to ensure that all relevant information is available in order to develop a product roadmap. Usually, this input is generated by involving customers or stakeholders. Typical examples for such information are the knowledge about the ecosystem, in which the company operates, the description of the problem that the product should solve, or the expected outcome of the solution.

2. *Establishing the 'why' with product vision and strategy:* This phase focuses on the formulation of the product vision and product strategy. This is important in order to align all stakeholders with the product roadmap and to obtain the agreement of the key stakeholders. A product vision should describe the "why" behind a product, i.e. the impact a product will have on the lives of people and the organization. Furthermore, the product vision is the justification for all subsequent efforts and forms the basis for the roadmap. If an organization has several products, the product vision is likely to be different from the corporate vision but should still be supportive and derived from it. The product strategy is the bridge that connects the product vision with the components of the roadmap and should describe how a product vision should be achieved.

3. *Uncovering customer needs:* This step is a very important aspect of the roadmapping process. The reason is that a suitable product roadmap for a dynamic and uncertain market environment should not only describe what will be build but also how it connects to the "why". This provides a clear direction that can be achieved in a number of ways. Usually, this connection is made by identifying the tasks that the customer has to fulfill or the problems that the customer has to solve. The components of the roadmap should be derived from these customer needs. One practice to uncover customer needs is the creation of "User Journey Maps". This involves analyzing the paths taken by customers through typical situations in which problems may arise (for example, online shopping).

4. *Deepening the product roadmap:* In this step, additional components can be added to the product roadmap according to the needs of the different stakeholders. The authors mention the following secondary components that can be added: 1) features and solutions (they should only be added for the short-term planning or for the internal communication purposes), 2) stage of development (e.g., "product discovery", "in development" or "testing"), 3) confidence (i.e., a percentage that indicates the probability of a roadmap component that it fulfils its expectations), 4) the target customer(s) and 5) the different product areas. For instance, the engineering teams can use product areas as an organizing principle for their internal teams. It should be noted that a product roadmap is valuable even without these components. One should be careful when adding secondary components, as they can increase complexity. Sometimes it makes more sense to create different documents instead of including all the details in the product roadmap. A product roadmap should not be a release plan.

5. *Prioritization:* The prioritization process should aim at finding the most efficient and effective way to deliver value to the customer and the business. A clear prioritization process helps to integrate all stakeholders' needs at an early stage and align these needs with the priorities.

6. *Achieve buy-in and alignment:* The product roadmap will not fulfill its purpose without alignment and buy-in of the key stakeholders. Examples for suitable approaches in order to gain alignment are the definition and communication of clear objectives based on the product vision or the conduction of shuttle diplomacy or mission briefing. More detailed processes and techniques can be found in Lombardo et al. [3].

7. *Presenting and sharing the roadmap:* Sharing the roadmap helps to obtain buy-in from a variety of stakeholders as well as from customers. It should be noted that different stakeholders require different kinds of information. For this reason, it is important to adjust the content of the presentation technique to the respective target group.

Another important aspect is that a product roadmap is subjected to frequent adjustments due to the dynamic market environment with high uncertainties. These adjustments should only be done in a systematic and justified way. Otherwise, this would result in a lot of rework and other negative consequences. Therefore, it is essential to establish a process for reassessing the product roadmap on a regular basis as well as to communicate the reason for changing the roadmap.

Holmes [25] suggests using a 3-step approach to create a roadmap which consists of the following phases: *1) collect input, 2) curate the inputs,* and *3) create the roadmap.* The phase *"collect inputs"* typically includes the collection of feedback (e.g., from external stakeholders or internal team members), the analysis of key metrics (e.g., to identify trends which help to build the roadmap), or the discussion of ideas in the product team. In order to generate ideas, the author recommends conducting a so-called "product council". A product council is a meeting with all key stakeholders to discuss different ideas and their priorities, and each stakeholder argues what they need and why. This approach fosters not only the communication between the product team and the stakeholders but also the communication and discussion between the stakeholders themselves. Before an idea is included into the product roadmap, it has to be evident that each idea contributes to achieving the product vision and goals and that the most valuable idea is implemented first. This is the starting point of the phase *"curate the inputs"*. In order to validate that each idea is aligned with the product vision and the objectives, the author recommends asking the following questions: 1) How does the product contribute to achieving the objectives? 2) What problem does the product solve? 3) What happens if we do not develop the product, and 4) what evidence exists that indicates the success of the product? Regarding prioritization, Holmes [25] proposes for instance an "Impact Effort Prioritization Matrix". For the phase *"creation of the product roadmap"*, the author recommends using the following guiding principles: 1) Keep it clear and simple: The primary purpose of the product roadmap it to communicate the strategic plan across the organization to stakeholders and if required to other parts of the business. If the roadmap is not simple and clear it is more difficult to communicate it. 2) Pick the right tool: Not every tool for product roadmapping is suitable for every team. For example, some teams like to work with a digital tool (e.g., Roadmunk, ProductPlan, or Trello), while some teams prefer to use a whiteboard. 3) Be disciplined: Regular updates of the roadmap should be made, otherwise it is not useful. This requires discipline in order to keep the roadmap updated and evolve it over time.

Van Os [26] recommends combining product roadmapping with design thinking. The author defines design thinking as framework for defining the problems that the end users are currently facing, which in turn uncovers potential opportunities for creating value. The process to create a product roadmap using the principles of design thinking consists of the following steps:

1) *Define the problems*: Talk to users in order to identify which problems cause them the most pain. The more pain a product can eliminate, the more valuable the user will find it. 2) *Determine the risks:* Prioritize the identified problems from the previous phase in order of importance. To conduct this step, the author recommends using the following questions: Is the money available to solve the problem or would it be better to spend the money elsewhere? Is there a team available to focus on the development of a solution in order to solve the problem? Is the problem in competition with other business objectives? Will people use what is being built and how likely are people willing to pay for the solution? 3) *Solutions and prototypes:* Determine possible solutions and create prototypes in order to gather feedback with respect to the current status of the proposed solution. This should be done by involving different teams (e.g., product management, engineering, sales) as different perspectives can be brought into the picture. Furthermore, more perspectives reduce the bias that could influence the proposed solution. The level of detail of the prototype should be based on the previously identified risks. Solutions with a high level of risk require detailed prototypes and more thorough testing than solutions with a low level of risk. In addition, when creating a prototype, care should be taken to ensure to create the minimum version of the solution required, as this limits the time wasted. 4) *Build the roadmap:* Prioritize the value and risk of each problem. Therefore, the author recommends using the value/risk mapping as shown in Fig. 4.

In order to make a decision on the different issues, the author recommends using the following guidelines:

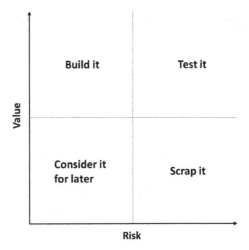

Fig. 4. Prioritization of roadmap items according to van Os [26]

- *High value/low risk:* Add the issue to the roadmap as they offer high value for the customer while the risk is low.
- *High value/high-risk:* Perform further testing to gather evidence about value creation.
- *Low value/low risk:* Leave the issues in consideration, but off the roadmap. The reason is that for the customers these issues are not important.
- *Low value/high risk:* Exclude these issues from the product roadmap as they provide only little value to the customer but are associated with a high risk.

Perri [27] proposes to consider product roadmaps as problem roadmaps (see Fig. 5). To create a problem roadmap Perri suggests the following 8-step approach: *1) Problem identification:* identify and list the problems customers or users are facing now; *2) Prioritization:* prioritize the list of problems with the strongest on top; *3) Assignment:* assign product teams to the problems to tackle it in a timeframe (e.g., one problem to one team in a quarter) and a KPI to measure their progress in solving the problem; *4) Problem validation:* each team should be responsible for validating its problems (i.e., to test if the problem exists and is worth solving). In the case that a problem turns out to be not relevant for the customers, the team can move on to the validation of the next problem on the list; *5) Development and validation of a Minimum-Viable Product:* after the initial user research and first validation results the teams will start to develop an MVP and test it in the market. It should be noted, that this step starts shortly after the previous step and runs mostly parallel; *6) Validation of the MVP:* Once a MVP has been validated, i.e. has met the goals, the next step is to start building a solution in the timeframe left. The focus is on minimum feature sets and on releasing often to get customer feedback as basis for further iterations; *7) Decision about the next steps:* based on the customer feedback the teams can decide whether more investment is required, regarding the evolvement of the current solution or whether another set of problems should be solved. At the beginning of the next iteration the process repeats; *8) Next iteration:* the prioritization of the list of problems should be revisited and potentially changed. New problems might be added.

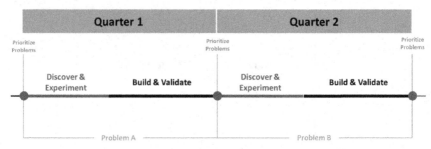

Fig. 5. Process to create a problem roadmap according to Perri [27].

The purpose of a problem roadmap is to support teams in changing their mindset from a feature-driven mindset to an outcome-oriented mindset. The problem roadmap consists of the two phases 1) discover and experiment, and 2) build and validate.

To compare the different product roadmap processes, Table 5 shows what the different approaches focus on.

Table 5. Important components of a roadmap process

Process component	Lombardo et al. [3]	Holmes [25]	Van Os [26]	Perri [27]
Achieving alignment and buy-in	x	x		
Gathering inputs from stakeholders	x	x		
Establishing the product vision and strategy	x			
Setting goals	x		x	x
Uncovering and selecting customer needs	x		x	x
Discovering and experimenting			x	x
Prioritizing	x	x	x	x
MVP creation and solution building			x	x
Assessing the contribution to the customer and business goals	x	x		x

5   Threats to Validity

We use the framework from Wohlin et al. [28] as the basis for the discussion of the validity of our study. **Construct validity:** The construct validity considers to what extent the operational measures represent what is investigated in the context to the research question [29]. First the construct validity is threatened by the Google search engine regarding the accessibility of search results. This means that Google does not allow the access to all identified articles by the search engine itself. Therefore, it is not known whether the articles returned by Google is representative of the total population of search results. Moreover, the search string itself poses a threat to the construct validity. There may be articles that deal with product roadmapping but use terms that were not covered by our search string. Therefore, we may have missed some relevant articles. In addition, a bias regarding the search history due to the Google's identify tracking mechanism cannot completely be excluded. This means that Google may return slightly different results if the search has been conducted repeatedly with our search string and the application of the same measures (conduction of the search in the incognito mode of the browser, usage of a VPN service, application of the relevance ranking and the activation of the option "inclusion of similar results"). **Internal validity:** The internal validity concerns the validity of the methods used to examine and analyze the data. To mitigate this threat, the quality assessment was conducted by two reviewers independently to limit confirmation bias and interpretation bias. In the case that the individual reviews led to different results, the process was repeated by a third reviewer to make a final decision. **External Validity:** The external validity considers to what extend it is possible to generalize the findings. The results and conclusion relate to product roadmapping in a dynamic market environment with high uncertainties (e.g., the software-intensive business). Therefore, the results are not directly transferable to other industry sectors. **Conclusion validity:** The validity of conclusions concerns the degree to which the conclusions of a study are based on the available data. To mitigate this risk, we have presented and discussed our findings with practitioners of the software-intensive business. In this context, no major ambiguities or inconsistencies were found [28, 29].

6 Summary

In this study, we conducted a review of the grey literature and identified processes suitable for the development and implementation of product roadmaps in dynamic and uncertain market environments. Overall, we identified 32 approaches that are similar in sequence and content. Therefore, we presented four representative roadmapping processes in this paper. An important common characteristic of the roadmapping processes found is that they establish a link between the "why" (i.e., the product vision and/or goals) and the "what" (i.e., the items to be delivered) of a product roadmap. The processes also focus on defining business and customer goals and keep how these goals can be achieved largely flexible. Finally, the processes support the validation of the items (i.e., through discovery, experimentation, prototyping, or assessing of the contribution to the goals). This should ideally be done before roadmap items are implemented, so that waste by developing the wrong items can be avoided and the odds of product success can be raised.

Appendix. Articles Identified by the Grey Literature Review

I.   280Group: What is a Product Roadmap?, available at https://280group.com/
        what-is-product-management/documents-templates/product-roadmap/,   last
        accessed 1st October 2020.

II.   Abate, S.: A Guide to Assembling a Product Roadmap, available at
        https://www.productplan.com/assembling-a-product-roadmap/, last accessed
        1st October 2020.

III.   Aha!: How Do Product Managers Build an Agile Roadmap?, avail-
        able at https://www.aha.io/roadmapping/guide/product-roadmap/how-do-pro
        duct-managers-build-an-agile-roadmap, last accessed 1st October 2020.

IV.   Blueprint: Agile Product Roadmap, available at https://www.blueprintsys.
        com/agile-planning/agile-product-roadmap, last accessed 1st October 2020.

V.   Choudhary, N.: Best Practices for Creating a Compelling Product Roadmap,
        available at https://www.tothenew.com/blog/best-practices-for-creating-a-
        compelling-product-roadmap/, last accessed 1st October 2020.

VI.   Holmes, R: Product roadmaps – an essential guide, available at https://
        www.departme tofproduct.com/blog/product-roadmaps-guide/, last accessed
        1st October 2020.

VII.   Justinmind: What is an agile roadmap? Product development tips, available
        at https://blog.prototypr.io/what-is-an-agile-roadmap-product-development-
        tips-3d81fcb59c88, last accessed 1st October 2020.

VIII.   Kukhnavets, P.: How to Make a Product Roadmap from Scratch?, avail-
        able at https://hygger.io/blog/make-product-roadmap-scratch/, last accessed
        1st October 2020.

IX.   Layton, M. C.: Four Steps to Creating an Agile Product Roadmap, available
        at https://www.dummies.com/careers/project-management/four-steps-to-cre
        ating-an-agile-product-roadmap/, last accessed 1st October 2020.

X.   Mitra, M.: Don't let your roadmapping process put you in handcuffs, available
        at https://www.productboard.com/blog/improve-your-product-roadmapping-
        process/, last accessed 1st October 2020.

XI.   Perri, M.: Rethinking the product roadmap, available at https://melissape
        rri.com/blog/2014/05/19/rethinking-the-product-roadmap, last accessed 1st
        October 2020.

XII.   Pichler, R.: Working with the GO Product Roadmap, available at https://
        www.romanpichler.com/blog/working-go-product-roadmap/, last accessed
        1st October 2020.

XIII.   ProductPlan: What is a Software Roadmap?, available at https://www.produc
        tplan.com/what-is-a-software-roadmap/, last accessed 1st October 2020.

XIV.   Product Manager HQ: How To Build A Product Roadmap Everyone Under-
        stands, available at https://productmanagerhq.com/how-to-build-a-product-
        roadmap-everyone-understands/, last accessed 1st October 2020.

XV.   Lombardo, C. T., McCarthy, B., Ryan, E., Conners, M.: Product roadmaps
        relaunched - How to set direction while embracing uncertainty. O'Reilly
        Media, Inc., Gravenstein Highway North, Sebastopol, CA, USA (2017).

XVI.   Melone, J.: How to successfully marry design sprints and product development, available at https://www.invisionapp.com/inside-design/how-to-suc cessfully-marry-design-sprints-and-product-development/, last accessed 1st October 2020.

XVII.   Roadmunk: A 5-step guide to agile product roadmap planning, available at https://roadmunk.com/guides/a-5-step-guide-to-agile-product-roadmap-planning/, last accessed 1st October 2020.

XVIII. Rai, A.: How to Build a Product Roadmap, available at https://hacker noon.com/how-to-build-a-product-roadmap-76b538923c6f, last accessed 1st October 2020.

XIX.   Seet, C.: Developing a strategy roadmap with design thinking, lean startup and agile, available at https://www.jibility.com/design-thinking-lean-startup-agile-strategy-roadmap/, last accessed 1st October 2020.

XX.   Sharma, R.: How to Craft the Perfect Product Roadmap, available at https://blog.hubspot.com/service/product-roadmap, last accessed 1st October 2020.

XXI.   Shymansky, S.: Product Development Roadmap – Your Guide Through the Product Strategy, available at https://railsware.com/blog/product-roadmap/, last accessed 1st October 2020.

XXII.   Smartsheet: Best Practices and Expert Tips for Creating Product Roadmaps, available at https://www.smartsheet.com/best-practices-and-expert-tips-cre ating-product-roadmaps, last accessed 1st October 2020.

XXIII. Svitla: Revealing the secrets of creating an efficient Agile Roadmap, available at https://svitla.com/blog/revealing-the-secrets-of-creating-an-efficient-agile-roadmap, last accessed 1st October 2020.

XXIV.   Tech at GSA: 3 Steps to Develop an Agile Product Roadmap, available at https://tech.gsa.gov/guides/develop_an_agile_product_roadmap/, last accessed 1st October 2020.

XXV.   Theus, A.: Building Your First Product Roadmap from Scratch, available at https://www.productplan.com/building-your-first-product-roadmap/, last accessed 1st October 2020.

XXVI.   Trackmind.com: Creating a strategic product roadmap, available at https://www.trackmind.com/creating-strategic-product-roadmap/, last accessed 1st October 2020.

XXVII. Tsui, W.: How to use design sprints to push your company roadmap, available at https://medium.com/gusto-design/how-to-use-design-sprints-to-push-your-company-roadmap-f085bc9d9a7c, last accessed 1st October 2020.

XXVIII. Umbach, H.: 8 Steps to a Successful Product Roadmap, available at https://www.freshtilledsoil.com/8-steps-to-a-successful-product-roa dmap/, last accessed 1st October 2020.

XXIX.   Van Os: The art of the strategic product roadmap, available at https://produc tcoalition.com/the-art-of-the-strategic-product-roadmap-c881f261b4eb, last accessed 1st October 2020.

XXX.   Walton, M.: Creating Good Roadmaps: 6 Practical Steps for Product Leaders, available at https://www.mindtheproduct.com/creating-good-roadmaps-6-pra ctical-steps-product-leaders/, last accessed 1st October 2020.

XXXI.   Weiss, E.: Build the Right Things - A 5 Step Plan to a Rock-Solid Product Roadmap, available at https://hackernoon.com/build-the-right-things-a-5-step-plan-to-a-rock-solid-product-roadmap-23a7c3fbd8d7, last accessed 1st October 2020.

XXXII. Zinchenko, P.: Agile product roadmap: pros, cons, and best practices, available at https://www.mindk.com/blog/product-roadmap/, last accessed 1st October 2020.

References

1. Kostoff, R.N., Schaller, R.: Science and technology roadmaps. IEEE Trans. Eng. Manage. **48**(2), 132–143 (2001)
2. Cooper, R.G., Edge, J.: Developing a product innovation and technology strategy for your business. Res. Technol. Manag. **53**(3), 33–40 (2015)
3. Lombardo, C.T., McCarthy, B., Ryan, E., Conners, M.: Product Roadmaps Relaunched - How to Set Direction While Embracing Uncertainty. O'Reilly Media, Inc., Sebastopol (2017)
4. Münch, J., Trieflinger, S., Lang, D.: The product roadmap maturity model DEEP: validation of a method for assessing the product roadmap capabilities of organizations. In: Hyrynsalmi, S., Suoranta, M., Nguyen-Duc, A., Tyrväinen, P., Abrahamsson, P. (eds.) ICSOB 2019. LNBIP, vol. 370, pp. 97–113. Springer, Cham (2019). https://doi.org/10.1007/978-3-030-33742-1_9
5. Lehtola, L., Kauppinen, M., Kujala, S.: Linking the business view to requirements engineering long-term product planning by roadmapping. In: 13th IEEE International Conference on Requirements Engineering, pp. 439–443. IEEE (2005)
6. Pichler, R.: Establishing an effective product strategy process. https://dzone.com/articles/establishing-an-effective-product-strategy-process. Accessed 1 Oct 2020
7. Münch, J., Trieflinger, S., Lang, D.: What's hot in product roadmapping? Key practices and success factors. In: Franch, X., Männistö, T., Martínez-Fernández, S. (eds.) PROFES 2019. LNCS, vol. 11915, pp. 401–416. Springer, Cham (2019). https://doi.org/10.1007/978-3-030-35333-9_29
8. Münch, J., Trieflinger, S., Lang, D.: Why feature based roadmaps fail in rapidly changing markets: a qualitative survey. In: Proceedings of International Workshop on Software-Intensive Business: Start-Ups, Ecosystems and Platforms (SiBW 2018), pp. 202–2018. CEUR-WS (2018)
9. Bowen, G., Bowen, D.: Strategy formulation and uncertainty in environments. J. Bus. Econ. **5**(12), 2315–2326 (2014)
10. Trieflinger, S., Münch, J., Bogazköy, E., Eißler, P., Schneider, J., Roling, B.: Product roadmap alignment – achieving the vision together: a grey literature review. In: Paasivaara, M., Kruchten, P. (eds.) XP 2020. LNBIP, vol. 396, pp. 50–57. Springer, Cham (2020). https://doi.org/10.1007/978-3-030-58858-8_6
11. Münch, J., Trieflinger, S., Bogazköy, E., Eißler, P., Roling, B., Schneider, J.: Product roadmap formats for an uncertain future: a grey literature review. In: Proceedings of Euromicro Conference on Software Engineering and Advanced Applications (SEAA 2020), pp. 284–291. IEEE (2020)
12. Münch, J., Trieflinger, S., Lang, D.: From vision to reality: a systematic literature review. In: Proceedings of ICE/IEEE ITMC International Conference on Engineering, Technology and Innovation, Valbonne, France (2019)
13. Kappel, T.: Perspectives on roadmaps: how organizations talk about the future. J. Prod. Innov. Manag. Int. Publ. Prod. Dev. Manag. Assoc. **18**(1), 39–50 (2001)

14. DeGregorio, G.: Technology management via a set of dynamically linked roadmaps. In: Proceedings of the 2000 IEEE Engineering Management Society, EMS 2000 (Cat. No. 00CH37139), pp. 184–190. IEEE (2000)
15. Albright, R.E.: A unifying architecture for roadmaps frames a value scorecard. In: IEMC 2003 Proceedings. Managing Technologically Driven Organizations: The Human Side of Innovation and Change. IEEE (2003)
16. Phaal, R., Farrukh, C.J.P., Probert, D.R.: Technology roadmapping—a planning framework for evolution and revolution. Technol. Forecast. Soc. Chang. **71**(1–2), 5–26 (2004)
17. Groenveld, P.: Roadmapping integrates business and technology. Res. Technol. Manag. **40**(5), 48–55 (1997)
18. Vähäniitty, J., Lassenius, C., Rautiainen, K.: An approach to product roadmapping in small software product businesses. In: Proceedings of the 7th European Conference on Software Quality (ECSQ 2002), pp. 12–13. Springer, Heidelberg (2002)
19. Fenwick, D., Daim, T.U., Gerdsri, N.: Value driven technology road mapping (VTRM) process integrating decision making and marketing tools: case of Internet security technologies. Technol. Forecast. Soc. Chang. **76**(8), 1055–1077 (2009)
20. Beeton, D., Phaal, R., Probert, D.R.: Exploratory roadmapping: capturing, structuring and presenting innovation insights. In: Moehrle, M.G., Isenmann, R., Phaal, R. (eds.) Technology Roadmapping for Strategy and Innovation, pp. 225–240. Springer, Heidelberg (2013). https://doi.org/10.1007/978-3-642-33923-3_14
21. Kim, E., Chung, J., Beckman, S., Agogino, A.M.: Design roadmapping: a framework and case study on planning development of high-tech products in Silicon Valley. J. Mech. Des. **138**(10), 101106 (2016)
22. Garousi, V., Felderer, M., Mäntylä, M.V.: Guidelines for including grey literature and conducting multivocal literature reviews in software engineering. Inf. Softw. Technol. **106**, 101–121 (2019)
23. Published on Figshare. https://figshare.com/s/109e070bd713512da5b0. Accessed 17 Nov 2020
24. Trieflinger, S., Münch, J., Knoop, V., Lang, D.: Facing the challenges with product roadmaps in uncertain markets: experience from industry. In: Proceedings of International Conference on Engineering, Technology and Innovation (ICE/ITMC 2020), pp. 1–8. IEEE (2020)
25. Holmes, R.: Product roadmaps – an essential guide. https://www.departmen-tofproduct.com/blog/product-roadmaps-guide/. Accessed 1 Oct 2020
26. Van Os: The art of the strategic product roadmap. https://productcoali-tion.com/the-art-of-the-strategic-product-roadmap-c881f261b4eb. Accessed 1 Oct 2020
27. Perri, M.: Rethinking the product roadmap. https://melissaperri.com/blog/2014/05/19/rethin king-the-product-roadmap. Accessed 1 Oct 2020
28. Wohlin, C., Runeson, P., Höst, M., Ohlsson, M., Regnell, B., Wesslén, A.: Experimentation in Software Engineering: An Introduction. Kluwer Academic Publishers, Dordrecht (2000)
29. Runeson, P., Höst, M.: Guidelines for conducting and reporting case study research in software engineering. Empir. Softw. Eng. **14**(2), 131–164 (2009). https://doi.org/10.1007/s10664-008-9102-8

Experience vs Data: A Case for More Data-Informed Retrospective Activities

Christoph Matthies$^{(\boxtimes)}$ and Franziska Dobrigkeit

Hasso Plattner Institute, University of Potsdam, Potsdam, Germany
{christoph.matthies,franziska.dobrigkeit}@hpi.de

Abstract. Effective Retrospective meetings are vital for ensuring productive development processes because they provide the means for Agile software development teams to discuss and decide on future improvements of their collaboration. Retrospective agendas often include activities that encourage sharing ideas and motivate participants to discuss possible improvements. The outcomes of these activities steer the future directions of team dynamics and influence team happiness. However, few empirical evaluations of Retrospective activities are currently available. Additionally, most activities rely on team members experiences and neglect to take existing project data into account. With this paper we want to make a case for data-driven decision-making principles, which have largely been adopted in other business areas. Towards this goal we review existing retrospective activities and highlight activities that already use project data as well as activities that could be augmented to take advantage of additional, more subjective data sources. We conclude that data-driven decision-making principles, are advantageous, and yet underused, in modern Agile software development. Making use of project data in retrospective activities would strengthen this principle and is a viable approach as such data can support the teams in making decisions on process improvement.

Keywords: Retrospective · Scrum · Agile methods · Data-driven decision making · Data-informed processes

1 Introduction

Agile development methods, particularly Scrum, which focus on managing the collaboration of self-organizing, cross-functional teams working in iterations [1], have become standards in industry settings. The most recent survey of Agile industry practitioners by *Digital.ai*[1], conducted between August and December 2019, showed that Scrum continued to be the most widely-practiced Agile method: 75% of respondents employed Scrum or a Scrum hybrid [2]. In the survey, which included 1,121 full responses, both of the top two Agile techniques employed in organizations were focused on communication and gathering feedback: the Daily Standup (85%) and Retrospective meetings (81%). The

---

[1] Formerly *CollabNet VersionOne*.

© Springer Nature Switzerland AG 2021
A. Przybyłek et al. (Eds.): LASD 2021, LNBIP 408, pp. 130–144, 2021.
https://doi.org/10.1007/978-3-030-67084-9_8

importance of these meetings was also stated in a previous, similar survey by the *Scrum Alliance*. A vast majority of respondents (81%) to this 2018 survey stated that their teams held a Retrospective meeting at the end of every Sprint, while 87% used Daily Scrum meetings [3]. A prototypical, generalized flow through the Scrum method, depicting the different prescribed meetings and process artifacts, i.e. the context of Retrospectives, is represented in Fig. 1.

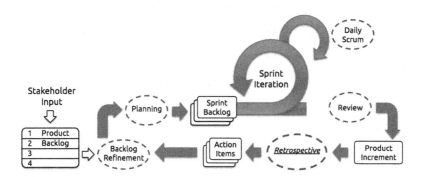

Fig. 1. Prototypical flow through the Scrum process, based on [4]. Process meetings are represented by circles, process artifacts and outcomes as squares. Scrum's process improvement meeting, the Retrospective, is highlighted.

In this research, we focus on the popular Retrospective meeting, which forms the core practice of process improvement approaches in the Scrum method, and the activities that are used in it. Retrospectives are a realization of the "inspect and adapt" principle [1] of Agile software development methods [5].

1.1   Retrospective Meetings

Recent research has pointed to Retrospective meetings as crucial infrastructure in Scrum [6]. Similarly, Retrospectives have also been recognized as one of the most important aspects of Agile development methods by practitioners [7]. The seminal work on Scrum, the *Scrum Guide*, defines the goal of Retrospectives to ascertain "how the last Sprint went" regarding both people doing the work, their relationships, the employed process, and the used tools [1]. As such, Retrospectives cover improvements of both technical and social/collaboration aspects. Teams are meant to improve their modes of collaboration and teamwork, thereby also increasing the enjoyment in future development iterations [8]. The Scrum framework prescribes Retrospectives at the end of each completed iteration. Teams are meant to generate a list of improvement opportunities, i.e. "action items" [8], to be tackled in the next iteration. Retrospective meetings focus less on the quality of the produced product increment, but more on how it was produced and how that process can be made smoother and more enjoyable for all involved parties in the next iteration.

While Scrum is a prescriptive process framework, suggesting concrete meetings, roles, and process outcomes, the Scrum Guide also points out: "Specific tactics for using the Scrum framework vary and are described elsewhere" [1]. For Retrospectives, this means that while the meeting's goal of identifying improvement opportunities is clear, the concrete steps that teams should follow are not and are up to the individual, self-organizing Scrum teams [9,10]. One of the easiest and most effective ways to generate the types of process insights that Retrospectives require is by relying on those most familiar with the teams' executed processes: team members themselves. Their views and perceptions of the previous, completed development iteration are, by definition, deeply relevant as inputs for process improvement activities. Furthermore, these data points are collectible with minimal overhead, e.g. by facilitating a brainstorming session in a Retrospective meeting, and they are strongly related to team satisfaction [11].

1.2 Data Sources Used in Retrospective Meetings

Most of the data that forms the basis of improvement decisions in current Retrospectives is, at present, based on the easily collectible perceptions of team members. However, modern software development practices and the continuing trend of more automated and integrated development tools have opened another avenue for accessing information on teams' executed process: their *project data* [12–14]. This project data includes information from systems used for such diverse purposes as version control (what was changed, why, when?), communication (what are other working on?), code review (feedback on changes), software builds (what is the testing status?), or static analysis (are standards met?). The data is already available, as modern software engineers continuously document their actions as part of their regular work [15,16]. The development processes of teams, their successes as well as their challenges, are "inscribed" into the produced software artifacts [17]. This type of information, which can be used in Retrospectives, in addition to the subjective assessments of team members, has been identified as "a gold-mine of actionable information" [18]. More comprehensive, thorough insights into teams' process states, drawn from activities that make use of both project data and team members' perceptions, can lead to even better results in Retrospectives [8].

1.3 Research Goals

In this research, we focus on the integration of project data sources into Agile Retrospective meetings. In particular, we investigate to which extent project data analyses are already provided for in Retrospective activities and how more of them could benefit from data-informed approaches in the future. We provide an overview of popular activities and review the types of data being employed to identify action items, i.e. possible improvements. We highlight those activities that already rely on software project data in their current descriptions as well as those that could be augmented to take advantage of the information provided by project data sources. We argue that the principles of data-driven decision making, which have already been adopted in many business areas [19], are suitable and conducive, yet underused, in the context of modern Agile process improvement.

2 Retrospective Activities

The core concept of Retrospectives is not unique to Scrum. These types of meetings, focusing on the improvement of executed process and collaboration strategies, have been employed since before Agile methods became popular. Similarly, team activities or "games" that meeting participants can play to keep sessions interesting and fresh have been used in Retrospectives since their inception [20]. These, usually time-boxed, activities are interactive and designed to encourage reflection and the exchange of ideas in teams. Derby and Larsen describe the purpose of Retrospective activities as to "help your team think together" [8]. Retrospective games have been shown to improve participants' creativity, involvement, and communication as well as make team members more comfortable participating in discussions [10]. The core idea is that meeting participants already have much of the information and knowledge needed for future process improvements, but a catalyst is needed to start the conversation.

In 2000, Norman L. Kerth published a collection of Retrospective activities [20]. Additional collections were published in the following years by different practitioners as well as researchers [21–24]. Table 1 presents an overview of the literature containing collections of Retrospective activities.

Table 1. Sources of Retrospective activities in literature.

Year	Reference	Name of reference
2006	[8]	Agile retrospectives - making good teams great
2006	[21]	Innovation games
2013	[22]	The retrospective handbook
2014	[23]	Getting value out of agile retrospectives
2015	[24]	Agile retrospective kickstarter
2015	[11]	Fun retrospectives
2018	[25]	Retromat: run great agile retrospectives!

A generalized meeting agenda for Retrospective was proposed in 2006 by Derby and Larsen. It features five consecutive phases: (i) *set the stage* (define the meeting goal and giving participants time to "arrive"), (ii) *gather data* (create a shared pool of information), (iii) *generate insight* (explore why things happened, identify patterns within the gathered data), (iv) *decide what to do* (create action plans for select issues), and (v) *close* (focus on appreciations and future Retrospective improvements) [8]. This plan has since established itself and has been accepted by other authors [25]. Retrospective activities remain an open area of investigation and continued learning, with current research further exploring the field [4, 26, 27].

While research articles and books offer extensive collection efforts regarding Retrospective activities, Agile practitioners rely on up-to-date web resources in their daily work, rather than regularly keeping up with research literature [27, 28]. The *Retromat*[2] [25] is a popular, comprehensive and often referenced [7, 27, 29], online repository of Retrospective activities for meeting agendas. It currently contains 140 different activities for five Retrospective meeting phases[3].

3   Review of Retrospective Activities

As the Retromat repository represents the currently best updated, most complete list of Retrospective activities in use by practitioners [27, 29, 30], we employ its database as the foundation of our review. Our research plan contains the following steps:

- Extract activities that provide or generate inputs for discussion in Retrospectives
- Identify the specific data points being collected
- Categorize data points by their origin
- Study those activities in detail which already (or are close to) taking project data into account

3.1   Activity Extraction

The Retromat, following Derby and Larsen's established model [8], features activities and games for the five Retrospective phases[4] *set the stage, gather data, generate insight, decide what to do* and *close the Retrospective.* As this research focuses on the types of gathered inputs employed for meetings, we initially collected all activities classified by the Retromat as suitable for the *gather data* phase. This meeting phase aims to help participants remember and reflect and is aimed at collecting the details of the last iteration, in order to establish a shared understanding within the team. We extracted 35 activities intended for the *gather data* from the Retromat repository. These activities are listed in Table 4 in the Appendix.

Additionally, we reviewed the Retromat activities prescribed for all other phases to ensure that we did not miss any activities that gathered data as part of their proceedings. These could have been classified under different phases, as data gathering and analysis steps are often intertwined, or because the activity's main focus is broader than data collection. This step yielded an additional four

---

[2] Available at https://retromat.org.

[3] https://retromat.org/blog/history-of-retromat/.

[4] https://retromat.org/blog/what-is-a-retrospective/.

activities that, at least partly, base their procedures on collected data: "3 for 1 - Opening" (assessments of iteration results and number of communications), "Last Retro's Actions Table" (collecting assessments of previous action items), "Who said it?" (collecting memorable quotes), and "Snow Mountain" (using the Scrum burndown chart)[5]. The first three of these were classified in the Retromat under the *set the stage* phase, the last as *generate insights*.

3.2   Identification of Retrospective Inputs

We analyzed the textual descriptions provided within the Retromat collection for each of the extracted activities of the previous research step. We manually tagged each of the activities with labels regarding the specific data points that are collected and used as inputs for the following actions. Many activity descriptions featured subsequent aggregation and synthesis actions, e.g. dot-voting or clustering, from which we abstracted. The generated short data labels describe the specific outcomes of the initial data acquisition within activities. Examples include "numerical ratings of performed meetings", "notes on what team members wish the team would learn", or "collection of all user stories handled during the iteration". Multiple activity descriptions contained mentions of physical representations of collected data points, which we generalized. For example, we consider "index cards" and "sticky notes" filled by meeting participants with their ideas to be instances of the more general "notes". The results of this tagging step are shown in Table 2.

3.3   Classification of Retrospective Data Sources

We categorized activities based on the origins of their data inputs, using the generated descriptions. We distinguish whether the gathered data is (i) drawn solely from team members' perceptions (no mention/reliance on software project data), (ii) is directly extracted from project data sources, or (iii) is ambiguous, i.e. could be drawn from either source, depending on team context and interpretation. We consider the term "project data" as an overarching collection of software artifacts. We follow Fernández et al.'s definition of the term "software artifacts" [31], in that we consider them "deliverables that are produced, modified, or used by a sequence of tasks that have value to a role". These artifacts are often subject to quality assurance and version control and have a specific type [31].

The vast majority, i.e. 86% (30 of 35), of proposed *gather data* activities in the Retromat collection make no mention of software project data and do not take advantage of it. It should be noted that most of these *gather data* activities are very similar in terms of the type of collected data. They tend to deal with team members' answers to varying prompts or imagined scenarios, aimed at starting discussions. Examples of such prompts include "mad, sad, glad", "start, stop, continue", "good, bad, ugly" or "proud and sorry". All of these activities

---

[5] https://retromat.org/en/?id=70-84-106-118.

Table 2. Overview of the types of inputs employed in the selected Retrospective activities. Activities above the divide are part of the *gather data* phase, those below are included after reviewing the activities of other phases. The categories of activities that gather data through specific prompts are italicized.

Shortened name	#	Type of activity input (regarding last iteration)
Activities from the *gather data* Retrospective phase		
Timeline	4	List of memorable/personally significant events
Analyze Stories	5	Collection of user stories handled during the iteration
Like to like	6	Notes on things to *start doing, keep doing* and *stop doing*
Mad Sad Glad	7	Notes on events when team members felt *mad, sad* or *glad*
Speedboat/Sailboat	19	Notes on what drove the team *forward* & what *kept it back*
Proud & Sorry	33	Notes of instances of *proud* and *sorry* moments
Self-Assessment	35	Assessments of team state regarding Agile checklist items
Mailbox	47	Reports of events or ideas collected during the iteration
Lean Coffee	51	List of topics team members wish to be discussed
Story Oscars	54	Physical representations of completed user stories
Expectations	62	Text on what team members expect from each other
Quartering	64	Collection of everything the team did during iteration
Appreciative Inquiry	65	Answers to positive questions, e.g. best thing that happened
Unspeakable	75	Text on the biggest unspoken taboo in the company
4 Ls	78	Notes on what was *loved, learned, lacked* & *longed for*
Value Streams	79	Drawing of a value stream map of a user story
Repeat & Avoid	80	Notes on what practices to *avoid* and which to *repeat*
Comm. Lines	86	Visualization of the ways information flows in the process
Satisfaction Hist.	87	Numerical (1–5) ratings of performed meetings
Retro Wedding	89	Notes on categories something *old, new, borrowed* & *blue*
Shaping Words	93	Short stories on iteration, including a 'shaping word'
#tweetmysprint	97	Short texts/tweets commenting on the iteration
Laundry Day	98	Notes on *clean* (clear) & *dirty* (unclear/confusing) items
Movie Critic	110	Notes on movie critic-style categories: *Genre, Theme, Twist, Ending, Expected?, Highlight, Recommend?*
Genie in a Bottle	116	Notes on 3 wishes: for *yourself, your team* and *all people*
Hit the Headlines	119	Short headlines on newsworthy aspects of the iteration
Good, Bad & Ugly	121	Notes on categories *good, bad* & *ugly* concerning the iteration
Focus Principle	123	Assessments on relative importance of Agile Manifesto principles
I like, I wish	126	Notes on *likes* and *wishes* concerning the iteration
Delay Display	127	Notes on team *destination, delay* & *announcement*
Learning Wish List	128	Text on what team members wish the team would learn
Tell me something I don't know	133	Facts and questions, in game show fashion, on something that only one team member knows and most others do not
Avoid Waste	135	Notes on the *7 categories of waste* in the process
Dare, Care, Share	137	Notes on *bold wishes, worries* & *feedback/news*
Room Service	139	Notes on the prompts *Our work space helps me/us...* and *Our work space makes it hard to...*
Activities from phases *set the stage* and *generate insights*		
3 for 1	70	Points in coordinate plane of satisfaction with results and communication
Retro Actions Table	84	List of last Retrospectives action items
Who said it?	106	Quotes collected from project artifacts
Snow Mountain	118	Burndown chart of problematic Sprint

are, by default, drawn from the individual perceptions and experiences of team members.

The nine activities that we identified in our review as featuring (possible) connections to development data—five from the *gather data*, three from *set the stage* and a single one from the *generate insights* phase—are shown in Table 3 and are discussed in the following two sections.

Table 3. Overview of Retromat activities not solely reliant on team members' perceptions. Activities which could be connected to project data, depending on how they are executed, are marked as *Possible*.

#	Activity Name	Data used as (partial) input for the activity and subsequent steps	Project data
5	Analyze Stories	Collection of all user stories handled during the iteration	*Yes*
54	Story Oscars	Physical representation of all stories completed in the last iteration	*Yes*
84	Last Retro's Actions Table	List of outcomes of the last Retrospective, i.e. action items/improvement plans	*Yes*
106	Who said it?	Literal quotes of team members extracted from communication channels, e.g. emails, chat logs or ticket discussions	*Yes*
35	Agile Self-Assessment	Assessments of team state regarding Agile checklist items	*Possible*
64	Quartering-identify boring stories	Collection of "everything" the team did in the last iteration	*Possible*
70	3 for 1	Number of times team members coordinated in the last iteration	*Possible*
79	Value Stream Mapping	Drawing of a value stream map concerning a particular user story	*Possible*
118	Snow Mountain	The shape of the Scrum Burndown chart of a problematic iteration	*Possible*

4 Activities Already Reliant on Project Data

Of the overall nine activities identified in this research that feature (possible) connections to project data, four make direct mentions of specific development artifacts in their descriptions on Retromat:

– *Analyze Stories*
– *Story Oscars*
– *Last Retro's Actions Table*
– *Who said it?*

These are marked as *Yes* regarding the use of project data in Table 4. Of these four activities, two employ the user stories of the last iteration as inputs, which are analyzed and graded by meeting participants in the following steps. The other two are concerned with the outcomes of the last Retrospective meeting and an extract of intra-team communications. The user stories/work items of modern Agile teams are usually contained in an issue tracker system [32] or can be acquired in printed form from a shared workspace or board [7]. Persisting the outcomes of Retrospectives, i.e. making note of the resulting action items and documenting meeting notes, is a common practice of Agile processes [6] and enables the tracking of progress towards these goals. Furthermore, digital communication tools, e.g. bug reports, mailing lists, or online forums, and the artifacts that result from their usage form a core part of modern software development [33]. The fact that these project artifacts are already present and are produced as part of the regular tasks of modern software developers, means that they can be collected with minimal overhead [34].

The four Retrospective activities we identified in this review as already employing project data represent only a small fraction of the 140 overall activities included in the Retromat. However, these are the activities that explicitly follow Derby and Larsen's principle of having the *gather data* phase of Retrospective meetings "start with the hard data" [8]. The authors consider this "hard data" to include iteration events, collected metrics, and completed features or user stories. They point out that while it "may seem silly to gather data for an iteration that lasted a week or two", being absent for a single day of a week-long iteration already results in missing 20% of events. As such, reflecting on the completed iteration through the lens of project data can ensure a more complete overview for all team members. Furthermore, even when nothing was missed through absence, perceptions of iteration events vary between observers and different people exhibit different perspectives and understandings regarding the same occurrences [8]. Lastly, by focusing on project data, in addition to the "soft data" usually employed, teams can optimize their Retrospective meetings. The roles in teams tasked with facilitating Retrospectives are able to prepare the inputs for meeting activities beforehand, without relying on the presence of others. Team members are able to focus their attention on interpreting data instead of trying to remember the details of the last iteration. The time gained by reviewing, e.g. an already existing list of user stories rather than having to reconstruct it collaboratively, frees up more time for the actual Retrospective work of reflecting on process improvements using Retrospective activities.

5   Towards Data-Informed Retrospective Activities

The activities that we identified, depending on interpretation and context, as having a *possible* connection to project data, i.e. depending on concrete execution in teams, are "Agile Self-Assessment", "Quartering - Identify boring stories", "3 for 1", "Value Stream Mapping" and "Snow Mountain", see Table 3. In the following paragraphs, we discuss these activities and their relations with software project data in detail.

Agile Self-Assessment involves assessments of team members regarding the state of their own team, based on a checklist of items. Depending on the employed checklist, these assessments might involve quantifiable measurements, e.g. "time from pushing code changes until feedback from a test is received"[6] or can rely on entirely team members' perceptions, e.g. "the team delivers what the business needs most"[7]. By switching to a checklist featuring measurements based on Agile practice usage and project data [35], this activity can be modified to present a more objective, data-based process view.

---

[6] https://finding-marbles.com/2011/09/30/assess-your-agile-engineering-practices/.

[7] https://www.crisp.se/gratis-material-och-guider/scrum-checklist.

Quartering - Identify boring stories assumes a collection of "everything a team did" in the last iteration. The activity's description does not mention how this overview is achieved or how the data points are collected. By brainstorming all their activities, this overview can be collaboratively reconstructed from the memories of participants. Relying on project data could significantly speed up this (error-prone) method of data collection. Dashboards featuring all interactions with the version control system by team members, e.g. using GitHub[8], can present activity audits with minimal overhead, leaving more time in Retrospectives for discussion. Furthermore, the goal of quartering is to identify boring stories. While the "boringness" of a story/work item is, by definition, in the eye of the beholder, data from project issue trackers could provide an additional level of analysis: Stories with no discussion that were closed rapidly, needing only a few commits by a single author, might be ideal candidates to be discussed for this Retrospective activity.

3 for 1 combines, as the name suggests, the assessments of meeting participants regarding three categories: iteration results, team communication, and mood. Team members are asked to mark their spot in a coordinate plane using the axes "satisfaction with iteration result" and "number of times we coordinated" with an emoticon representing their mood. While satisfaction with iteration results and mood are hard to gauge using project data, the frequency of communication within a team can be extracted from the team's employed communication tools. As more communication moves to digital tools, such as chat or ticket systems, the wealth of information in this domain is steadily increasing [36]. If a digital tool is used, the number of contacts and touch points between team members can be counted and quantified. The input for one axis of the *3 for 1* activity can therefore be automated or augmented with project data analyses. Furthermore, variations of this exercise include varying the employed categories, such as replacing communication frequency with the frequency of pair programming [37] in the team. Relying more heavily on project data analyses for this activity can simplify both data collection and substitution of employed categories.

Value Stream Mapping attempts to create a *value stream map* (VSM) [38,39] of a team's process based on the perspective of a single user story. While the details of the story might still be in participants' memories, gathering additional data, based on project artifacts, can provide additional context to improve the map's accuracy. One of the main goals of a VSM is to identify delays, choke points, and bottlenecks in the process. In a software development process, these are measurable using project data, e.g. by calculating the time it took from pushing code for a story until the code was reviewed or by assessing its *lead time* [40]. A more complete VSM can be generated by relying on these metrics, leading to improved subsequent analysis and improvement steps in a team.

Snow Mountain uses the shape of the Scrum Burndown chart regarding a problematic iteration to draw an image that is used as a reflection prompt. Using the

---

[8] https://github.blog/changelog/2018-08-24-profile-activity-overview/.

metaphor of a snowy mountain ridge, meeting participants describe their perceptions of the iteration with kids sledging down the slopes. The Burndown chart is a measurement tool for planning and monitoring of progress in Scrum teams [41] They are based on the amount of work left to do versus remaining time during an iteration. Depending on the team, the amount of outstanding work can be represented by time units, story points or other effort measures (e.g. "gummy bear" [42]). If sophisticated project management software is used by the team and work items are entered into it with the required level of detail, burndown charts can be created and extracted from the project data[9]. These digital images can then be printed or otherwise transformed into the snowy mountains required for the activity, without expending team members' time in creating them.

6 Conclusion

We present a review and analysis of current Retrospective activities, with a focus on the *gather data* meeting phase. We discuss the role of software project data, i.e. development artifacts produced by developers in their day-to-day work, within existing Retrospective meeting structures. This type of data has previously been identified in the literature as an extremely valuable source of insight and actionable information [18,34]. However, we show that the vast majority, i.e. 86%, of activities explicitly proposed for the *gather data* phase in a popular Retrospective agenda collection [25], lack explicit connections to this software project data. Of these data-gathering activities, many share a similar process of collecting participant perceptions and improvement ideas through structured prompts in the general form of *start, stop, continue*. Most current Retrospective activities rely on the perceptions of meeting participants as their sole inputs. However, software project data, in particular requirements information or insights from version control systems, show promise as additional data sources for Retrospective techniques. Integrating the principles of data-driven decision-making, based on project data, into Agile processes enables "evidence-based decision making" [38] in Retrospective meetings.

These concepts are not foreign (or new) to Agile methods but seem to have fallen by the wayside recently. The Scrum Guide states, "Scrum is founded on empirical process control theory [...] knowledge comes from experience and making decisions based on what is known. [...]" [1]. We argue that these concepts are still important, yet underused, in current implementations of Agile methods in general and Retrospectives in particular.

We identify four meeting activities in the Retromat collection that already explicitly take project data into consideration. Of these, only two are listed for the *gather data* phase of Retrospectives. We then focus on employing software project data in additional activities to augment Retrospective meetings, decreasing manual efforts by Agile development teams, and process facilitators. We propose modifications to five other activities, which are suited to take advantage of

---

[9] https://support.atlassian.com/jira-software-cloud/docs/view-and-understand-the-burndown-chart/.

the process knowledge contained within project data. These proposals present initial steps towards more evidence-based, data-informed decision making by participants of Retrospectives.

Appendix

Table 4. List of activities extracted from the Retromat repository [25] for the *gather data* phase of Retrospectives, as of Oct. 2020.

#	Name & Activity Tagline
4	*Timeline:* Write down significant events and order them chronologically
5	*Analyze stories:* Walk through a team's stories and look for possible improvements
6	*Like to like:* Match quality cards to their own Start-Stop-Continue-proposals
7	*Mad sad glad:* Collect events of feeling mad, sad, or glad and find the sources
19	*Speedboat/Sailboat:* Analyze what forces push you forward and pull you back
33	*Proud & sorry:* What are team members proud or sorry about?
35	*Agile self-assessment:* Assess where you are standing with a checklist
47	*Empty the mailbox:* Look at notes collected during the iteration
51	*Lean coffee:* Use the Lean Coffee format for a focused discussion of the top topics
54	*Story oscars:* The team nominates stories for awards and reflects on the winners
62	*Expectations:* What can others expect of you? What can you expect of them?
64	*Quartering:* Categorize stories in 2 dimensions to identify boring ones
65	*Appreciative Inquiry:* Lift everyone's spirit with positive questions
75	*Writing the unspeakable:* Write down what you can never ever say out loud
78	*4 Ls:* Explore what people loved, learned, lacked and longed for individually
79	*Value stream mapping:* Draw a value stream map of your iteration process
80	*Repeat & avoid:* Brainstorm what to repeat and what behaviours to avoid
86	*Lines of communication:* Visualize information flows in, out and around the team
87	*Meeting satisfaction histogram:* Create a histogram on how well ritual meetings went during the iteration
89	*Retro wedding:* Collect examples for something old, new, borrowed and blue
93	*Tell a story with shaping words:* Each participant tells a story about the last iteration that contains certain words
97	*#tweetmysprint:* Produce the team's twitter timeline for the iteration
98	*Laundry day:* Which things are clear & feel good and which feel vague & implicit?
110	*Movie critic:* Imagine your last iteration was a movie and write a review about it
116	*Genie in a bottle:* Playfully explore unmet needs
119	*Hit the headlines:* Which sprint events were newsworthy?
121	*The good, the Bad, and the ugly:* Collect what team members perceived as good, bad and non-optimal
123	*Find your focus principle:* Discuss the 12 agile principles & pick one to work on
126	*I like, i wish:* Give positive, as well as non-threatening, constructive feedback
127	*Delay display:* What's the current delay? And where are we going again?
128	*Learning wish list:* Create a list of learning objectives for the team
133	*Tell me something I don't know:* Reveal hidden knowledge with a game show
135	*Avoid waste:* Tackle the 7 Wastes of Software Development
137	*Dare, care, share:* Collect topics in three categories: 'Dare', 'Care' and 'Share'
139	*Room service:* Take a look at the team room: Does it help or hinder?

References

1. Schwaber, K., Sutherland, J.: The scrum guide - the definitive guide to scrum: the rules of the game. Technical report, scrumguides.org (2017). http://scrumguides.org/docs/scrumguide/v2017/2017-Scrum-Guide-US.pdf
2. Digital.ai (formerly CollabNet VersionOne). 14th Annual State of Agile Report. Technical report (2020). https://explore.digital.ai/state-of-agile/14th-annual-state-of-agile-report
3. Scrum Alliance. State of Scrum 2017-2018: Scaling and Agile Transformation. Technical report (2018). http://info.scrumalliance.org/State-of-Scrum-2017-18.html
4. Matthies, C., Dobrigkeit, F.: Towards empirically validated remedies for scrum retrospective headaches. In: Proceedings of the 53rd Hawaii International Conference on System Sciences (2020). http://hdl.handle.net/10125/64504
5. Andriyani, Y., Hoda, R., Amor, R.: Reflection in agile retrospectives. In: Baumeister, H., Lichter, H., Riebisch, M. (eds.) XP 2017. LNBIP, vol. 283, pp. 3–19. Springer, Cham (2017). https://doi.org/10.1007/978-3-319-57633-6_1
6. Dingsøyr, T., Mikalsen, M., Solem, A., Vestues, K.: Learning in the large - an exploratory study of retrospectives in large-scale agile development. In: Garbajosa, J., Wang, X., Aguiar, A. (eds.) XP 2018. LNBIP, vol. 314, pp. 191–198. Springer, Cham (2018). https://doi.org/10.1007/978-3-319-91602-6_13
7. Kniberg, H.: Scrum and XP From the Trenches, vol. 2. C4Media, Toronto (2015). ISBN 9781430322641
8. Esther, D., Larsen, D.: Agile Retrospectives: Making Good Teams Great. Pragmatic Bookshelf (2006). ISBN 0-9776166-4-9
9. Matthies, C., Kowark, T., Uflacker, M.: Teaching agile the agile way — employing self-organizing teams in a university software engineering course. In: American Society for Engineering Education (ASEE) International Forum, ASEE (2016). https://peer.asee.org/27259
10. Przybyłek, A., Kotecka, D.: Making agile retrospectives more awesome. In: Proceedings of the 2017 Federated Conference on Computer Science and Information Systems, vol. 11, pp. 1211-1216 (2017). https://doi.org/10.15439/2017F423. ISBN 9788394625375
11. Caroli, P., Caetano, T.: Fun retrospectives-activities and ideas for making agile retrospectives more engaging. Leanpub.com (2016). https://leanpub.com/funretrospectives
12. Matthies, C., Dobrigkeit, F., Hesse, G.: Mining for process improvements: analyzing software repositories in agile retrospectives. In: Proceedings of the IEEE/ACM 42nd International Conference on Software Engineering Workshops, pp. 189-190. ACM (2020). https://doi.org/10.1145/3387940.3392168. ISBN 9781450379632
13. Zaitsev, A., Gal, U., Tan, B.: Coordination artifacts in agile software development. Inf. Organ. **30**(2), 100288 (2020). https://doi.org/10.1016/.infoandorg.2020.100288. ISSN 14717727
14. Wohlrab, R.: Living boundary objects to support agile inter-team coordination at scale. Ph.D. thesis (2020).https://research.chalmers.se/en/publication/515968
15. Ying, A.T. T., Wright, J.L., Abrams, S.: Source code that talks: an exploration of eclipse task comments and their implication to repository mining. ACM SIGSOFT Softw. Eng. Notes **30**(4), 1 (2005). https://doi.org/10.1145/1082983.1083152. ISSN 0163-5948

16. Matthies, C., Kowark, T., Richly, S., Uacker, M., Plattner, H.: ScrumLint: identifying violations of agile practices using development artifacts. In: Proceedings of the 9th International Workshop on Cooperative and Human Aspects of Software Engineering, pp. 40-43. ACM (2016). https://doi.org/10.1145/2897586.2897602. ISBN 9781450341554

17. de Souza, C., Froehlich, J., Dourish, P.: Seeking the source: software source code as a social and technical artifact. In: Proceedings of the 2005 International ACM SIGGROUP Conference on Supporting Group Work, pp. 197. ACM Press (2005). https://doi.org/10.1145/1099203.1099239. ISBN 1595932232

18. Guo, J., Rahimi, M., Cleland-Huang, J., Rasin, A., Hayes, J.H., Vierhauser, M.: Cold-start software analytics. In: Proceedings of the 13th International Workshop on Mining Software Repositories, pp. 142-153. ACM Press (2016). https://doi.org/10.1145/2901739.2901740. ISBN 9781450341868

19. Matthies, C., Hesse, G.: Towards using data to inform decisions in agile software development: views of available data. In: Proceedings of the 14th International Conference on Software Technologies, pp. 552-559. SciTePress (2019). https://doi.org/10.5220/0007967905520559. ISBN 978-989-758-379-7

20. Kerth, N.L.: The ritual of retrospectives: how to maximize group learning by understanding past projects. Softw. Test. Qual. Eng. **2**(5), 53-57 (2000)

21. Hohmann, L.: Innovation Games: Creating Breakthrough Products through Collaborative Play. Addison-Wesley, Boston (2006)

22. Kua, P.: The Retrospective Handbook: A Guide for Agile Teams. Leanpub.com (2013). https://leanpub.com/the-retrospective-handbook. ISBN 78-1480247871

23. Gonçalves, L., Linders, B.: Getting Value out of Agile Retrospectives - A Toolbox of Retrospective Exercises. Leanpub.com (2014). https://www.infoq.com/minibooks/agile-retrospectives-value/. ISBN 9781304789624

24. Krivitsky, A.: Agile Retrospective Kickstarter. Leanpub.com (2015). https://leanpub.com/agile-retrospective-kickstarter

25. Baldauf, C.: Retromat - Run great agile retrospectives! Leanpub.com (2018). https://leanpub.com/retromat-activities-for-agile-retrospectives

26. Jovanović, M., Mesquida, A.-L., Radaković, N., Mas, A.: Agile retrospective games for different team development phases. J. Univ. Comput. Sci. **22**(12), pp. 1489-1508 (2016). https://doi.org/10.3217/jucs-022-12-1489

27. Loeffer, M.: Improving Agile Retrospectives: Helping Teams Become More Efficient. Addison-Wesley Professional, Boston (2017). ISBN 978-0134678344

28. Beecham, S., O'Leary, P., Baker, S., Richardson, I., Noll, J.: Making software engineering research relevant. Computer **47**(4), 80-83 (2014). https://doi.org/10.1109/MC.2014.92. ISSN 0018-9162

29. Northwood, C.: Planning Your Work, pp. 11-46. Apress, New York (2018). https://doi.org/10.1007/978-1-4842-4152-3_2. ISBN 978-1-4842-4152-3

30. Stålesen, A.M., Dølvik, B.: Agile retrospectives: an empirical study of characteristics and organizational learning. Master thesis, Norwegian University of Science and Technology (2015)

31. Méndez Fernández, D., et al.: Artefacts in software engineering: what are they after all? (2018). http://arxiv.org/abs/1806.00098

32. Dimitrijević, S., Jovanović, J., Devedžić, V.: A comparative study of software tools for user story management. Inf. Softw. Tech. **57**(1), 352-368 (2015). https://doi.org/10.1016/j.infsof.2014.05.012. ISSN 09505849

33. Nazar, N., Hu, Y., Jiang, H.: Summarizing software artifacts: a literature review. J. Comput. Sci. Technol. **31**(5), 883-909 (2016). https://doi.org/10.1007/s11390-016-1671-1. ISSN 1000-9000

34. Ortu, M., Destefanis, G., Adams, B., Murgia, A., Marchesi, M., Tonelli, R.: The JIRA repository dataset. In: Proceedings of the 11th International Conference on Predictive Models and Data Analytics in Software Engineering, pp. 1-4. ACM Press (2015). https://doi.org/10.1145/2810146.2810147. ISBN 9781450337151

35. Matthies, C., Kowark, T., Uacker, M., Plattner, H.: Agile metrics for a university software engineering course. In: 2016 IEEE Frontiers in Education Conference, pp. 1-5. IEEE (2016). https://doi.org/10.1109/FIE.2016.7757684. ISBN 978-1-5090-1790-4

36. Stray, V., Moe, N.B.: Understanding coordination in global software engineering: a mixed-methods study on the use of meetings and Slack. J. Syst. Softw. **170**, 110717 (2020). https://doi.org/10.1016/j.jss.2020.110717. ISSN 01641212

37. Kniberg, H.: Scrum and XP from the Trenches. C4Media (2007). ISBN 978-1-4303-2264-1

38. Fitzgerald, B., Musia, M., Stol, K.-J.: Evidence-based decision making in lean software project management. In: Companion Proceedings of the 36th International Conference on Software Engineering - ICSE Companion 2014, pp. 93-102, New York, USA (2014). ACM Press. https://doi.org/10.1145/2591062.2591190. ISBN 9781450327688

39. Kupiainen, F., Mäntylä, M.V., Itkonen, J.: Using metrics in agile and lean software development - a systematic literature review of industrial studies. Inf. Softw. Technol. **62**(1), 143-163 (2015). https://doi.org/10.1016/j.infsof.2015.02.005. ISSN 09505849

40. Ahmad, M.O, Markkula, J., Oivo, M.: Kanban in software development: a systematic literature review. In: 2013 39th Euromicro Conference on Software Engineering and Advanced Applications, pp. 9-16. IEEE (2013). https://doi.org/10.1109/SEAA.2013.28. ISBN 978-0-7695-5091-6

41. Scott, E., Pfahl, D.: Exploring the individual project progress of scrum software developers. In: Felderer, M., Méndez Fernández, D., Turhan, B., Kalinowski, M., Sarro, F., Winkler, D. (eds.) PROFES 2017. LNCS, vol. 10611, pp. 341–348. Springer, Cham (2017). https://doi.org/10.1007/978-3-319-69926-4_24

42. Meyer, B.: Agile! The Good, the Hype and the Ugly. Springer, Cham (2014). https://doi.org/10.1007/978-3-319-05155-0

Reducing the Uncertainty of Agile Software Development Using a Random Forest Classification Algorithm

Ewelina Wińska[1]([⊠]) [iD], Estera Kot[2] [iD], and Włodzimierz Dąbrowski[2] [iD]

[1] Polish-Japanese Academy of Information Technology, 02-008 Warsaw, Poland
ewelinawinska@pjwstk.edu.pl
[2] Warsaw University of Technology, 00-661 Warsaw, Poland

Abstract. Background: Companies operating in the software industry or those which rely on new technologies are facing a rising level of complexity in building products. To address these new circumstances, enterprises are investing more resources in modern approaches to software delivery, such as agile methodologies. Amongst these methodologies, relative effort estimation is widely adopted. Outcomes of the estimation process are often not predictable or reliable. **Aims**: The objective of this paper is to research the random forest classification algorithm's effectiveness for high-level effort estimating. **Method**: Authors are focusing on defining complexity factors that are treated as model features. In addition, the authors have empirically tested the proposed solution in a commercial environment. Besides these, authors have analyzed the effective impact of each complexity factor. The analysis was done on the set of seventy thousands of Jira work items. Observation has been made empirically across four major releases. **Results**: The results indicate that the empirical way of defining model features has a significant impact on effort estimation accuracy. During research, the authors have found several key factors that have a significant impact on model accuracy. Teams that are using agile techniques or methods for effort estimation can enhance planning outcomes with tools supporting high-level estimation. Finding out and fine-tuning such tools needs a structured process for finding the most significant key complexity factors. **Conclusion**: Usage of metrics such as effort estimations and their accuracy in the software development process in agile organizations could lead to more accurate planning and forecasting of project outcomes. Problems with planning on program level could also be actioned with a structured estimation framework, enhanced by modern tools such as classification models. We should remember that complexity is growing with scaling delivery structures within companies.

Keywords: Effort estimation · Decision tree classifier · Extra tree classifier · Random forest classifier · Relative estimation · Software effort estimation · Agile software development

© Springer Nature Switzerland AG 2021
A. Przybyłek et al. (Eds.): LASD 2021, LNBIP 408, pp. 145–155, 2021.
https://doi.org/10.1007/978-3-030-67084-9_9

1   Introduction

We are living in a fast-paced environment, where changes are not always predictable. To accommodate for this uncertainty, companies are turning their focus on Agile methodologies. Changes in product requirements are not the problem, but rather the unnecessary effort that is committed for functionalities that are not needed, which results in the waste of resources [14]. Companies are shifting from following a waterfall approach to software delivery with a specific implementation like PRINCE2^{TM}, in order to incorporate shorter iterations that can deliver product increments of high-priority features and decrease Time-to-Market Value [12]. Via collaboration between Product Owner and Development Team and frequent feedback cycles, Scrum tends to produce high-quality deliverables with high transparency over the project. The reason behind large corporations shifting their focus towards agile practices and changing ways of working is to increase capabilities to react fast enough in the new changing environment. Predicting the cost of development is a complex task that depends on a variety of factors.

Due to its complexity, predictive effort estimation of effort forecasting could be misleading [7]. Relative estimations are commonly used by mature Agile Teams. It is a process of estimating task completion by comparing them to previously completed work items. There is no mention of a time requirement, just that is more or less complex than others. A method consistent with estimation in units other than time avoids some of the pitfalls associated with guessing estimations: unwarranted precision, confusing estimates for commitments. The human brain is naturally hard-wired to work better with relative comparison, an inbuilt sense of something being relatively bigger or smaller than something else. This explains why the development teams are much more comfortable with relative estimation. They are aware that they do not have all the necessary information at the time of estimating, and so they don't feel confident to say how long a task will take.

Program managers or stakeholders often mistakenly take time estimates as commitments putting trust in team expertise which leads to the expectation that features will be completed at this time. Miranda [8] presents a chart with data suggesting that comparing complexity leads to more accurate judgments than "ad hoc" estimates. Vicinanaza et al., claims that experts are more accurate in relative estimations than in absolute. Study [10] shows that a sample of five managers, shown data from past projects. One at a time was asked to estimate their effort, and they were more accurate than estimates derived from traditional lines of code estimation used in COCOMO which is an example of software engineering cost estimation model.

Instead of forecasting the precise amount of effort or time needed to accomplish the project goals, practitioners should focus on relative estimation using story points. Traditional software teams give estimates in a time format: days, weeks, months. Many agile teams, however, have transitioned to story points. Story points rate the relative effort of work in a Fibonacci-like format: 0, 0.5, 1, 2, 3, 5, 8, 13, 20, 40, 100. It may sound counter-intuitive, but that abstraction is actually helpful because it pushes the team to make tougher decisions around the difficulty of work.

This leads to the hypothesis that instead of attempting precise forecasting of the time needed for delivering specific pieces of work, effort estimation could be resolved as a classification problem. Strong estimates are very important, especially in predictable

product development [1]. The effort needed to deliver business value is the most important cause that is affecting the budget, and product roadmap. To achieve better generalization, organizations are using relative effort estimation instead of trying to predict the
exact amount of time to deliver a whole product. Relative estimation is highly dependent on a given context but could lead to better outcomes than forecasting or predicting
needed effort, even for long term planning. Such an approach to the effort estimation
process could be resolved as a classification problem where relative measures would be
treated as several imbalanced classes and random forest.

For this research, the authors favor three decision tree techniques, based on empirical
experience. It is also worth mentioning that not only decision trees can resolve classification problems. Classification can be also resolved with a single hidden layer perceptron
neural network, support vector machines, or k-means clustering [2]. The authors have
researched three classifiers: decision tree classifier, extra tree classifier.

Decision tree classifier [3]. A decision tree algorithm could be explained as a logical
model. Following, decision tree classifier is a kind of model that resolves classification
problems with a decision based on predefined conditions and possible outcomes. The
decision tree, in general, could be used for both classification and regression [4] problems. A decision tree can be visualized as having a flow diagram like structure, where
each internal node is modeling a test against one of the predefined attributes. Each branch
of the classifier resolves an outcome of the test, and each leaf holds a class label.

Extra tree classifier, which is an ensemble learning method fundamentally based
on decision trees. Extra tree classifier, like random tree classifier, randomizes certain
decisions and subsets of data to minimize overfitting. This example of a tree-based
classifier essentially utilizes the randomizing of both attribute and cut-point choice while
splitting a tree node. In some cases, the model is building random trees whose structures
are independent of the output values of the learning sample. This could lead to a positive
outcome of the randomization process, and in general a better model performance [5].

Random forest is another ensemble learning-based [6] model that could be used
for both classification and regression problems. A model during the learning process is
building a number of decision trees. Classes generated by distinctive trees are selected
by the model throughout the training process. The ensemble model combines the results
from the different decision trees. It is also important to mention that the model, in general,
is able to obtain more accurate outcomes than the model based on single decision tree
models.

2   Related Work

Software effort estimation is not a new practice; however, it differs through the industries.
Based on experience, and with supporting research, authors would like to show how effort
estimation is important. In the HELENA (Hybrid dEveLopmENt Approaches) survey
researchers were studying software development on hybrid development approaches in
software development from regulated domains to emerging and innovative sectors. The
goal was to investigate what is the current state of the practice in software and system
development. Focusing on practices that are used in the practice or production environments, and how experts are combining them together, how such combinations were

developed over time, and how standards affect the development process. The HELENA survey has been designed as three-staged international research. The first stage was aimed at preparing the data collection and to test the survey instrument. The authors were focusing on four research [15, 16]. In [15] authors received responses from Swedish companies. With response rate at level 37% 513 responses were collected. What is interesting for us is that the survey was asking respondents which practices related to Agile methodologies they are using. Authors would like to focus on responses around planning and estimating software effort. While 80% of respondents claim that they always or often use release planning practice, 70% of respondents were using guess-based expert or team-based estimation, formal estimation was used in 58% of cases, and Velocity based planning practice was used only in 32% of cases. In [16] authors had conducted detailed industry analysis from a large-scale online survey among practitioners. The survey was grounded with 1467 data points from large-scale online surveys. Survey results ended with an overview of the practices, it appeared that 85% of practices have been using common agreement in hybrid development methods. What is interesting for us: whilst all respondents reported that they are planning releases, no one claimed Velocity based planning nor formal estimation. Expert estimation was reported when using a combination of Scrum, Lead Software Development, Iterative Development, Kanban, and Devops frameworks. Authors are having an interesting conclusion that satisfaction correlates to the usage of agile software development methods. Research has shown a very high satisfaction rate, both for companies and individual professionals, with very similar values relates to methodologies used. Companies report high satisfaction while professionals are not contented with ways of working. It was stated by the authors in [16] lack of user stories, story mapping, lack of estimation appears strongly related to low satisfaction. Considering all these together suggest satisfaction relates to concern for quality work, team cohesion, and support for tracking progress. In [1] authors were researching the precision and reliability of the estimation of the effort of software projects. They found that the quality of estimates plays a very important role in the management of software projects or building software products. In their paper, authors have introduced a new method based on machine learning which gives the estimation of the effort together with a confidence interval for it. In their method, researchers proposed to employ confidence intervals that are independent of a probability distribution. In [11] Magne Jorgensen and Martin Shepperd have identified 304 software cost estimation papers in 76 journals and classified those papers according to the research topics: estimation approach, research approach, study context, and data sets. Researchers were aiming to provide a foundation for the improvement of software estimation research through a systematic review of previous work. In [3] authors are identifying the effort estimation process as one of the most complex problems faced by the software industry. One of the findings in the paper is that software planning estimation of the effort is one of the most critical responsibilities which is crucial for successful product delivery. Researchers are also mentioning that it is necessary to have good effort estimation in order to propose a well-prepared project budget. The accuracy of the effort estimation of software projects is vital for the competitiveness of companies that are working in the software industry. For better more accurate forecasting or estimating the effort needed for software project delivery, it is important to select the correct software effort estimation techniques.

3 Company Landscape and Research Environment

Research has been made during the realization of a transformation project in a commercial environment. The project was realized in one of the biggest European investment banks with departments distributed around the world. The main goal of the project was to deliver two hundred of new services and capabilities within the self-service portal based on ServiceNow IT service management platform. Project life span was initially planned for a two-year horizon. One of the initial assumptions was that it is possible to deliver most of the new services with ServiceNow out of the box functionalities and without a big number of complicated customizations. During an initial couple of months, the Product Owner was focusing on preparing the product road map and defining the goal. Initially, effort relative estimation was based on small, medium, and high complexity ratings. This approach was not given enough insight and was not useful in terms of road map monitoring or high-level planning. The actual team velocity did not match the initial estimates based. It has emerged that estimates were far from being accurate and this has led to not reliable plans and forecasts. After a number of releases, program management has decided to move to story points as an effort relative estimation unit of measure. Together with the introduction of story points as effort relative estimation units of measure, it appeared that a structured approach to estimation process is needed. The main purpose behind having a dedicated tool for enhancing the estimation process is to bring more understanding to the complexity of requested features or in this particular case, service that is needed for a self-service portal. Initially after gathering the answers a simple algorithm with defined weights was able to assign a complexity rating, as shown in Table 1. The purpose of the program development team was to maintain a library of reusable components. With such an approach, it was possible to address the increase of complexity over time.

Table 1. Table shows complexity used for high level estimation of features divided into five complexity classes.

Complexity	Rate = effort estimate in Story points
Very Low	50–99
Low	100–149
Medium	150–199
High	200–249
Very High	250–300

Initially the most common complexity rating was very high or high complexity. But gradually, together with better understanding of dependencies between new requirements analysis of work items complexity have been approached with higher attention to details. Responsible analysts were putting more effort into breaking down complex items into more independent work items. Together with this approach, the number of low complexity items has started to grow and at the end of the experiment, has stabilized around 40% of items estimated with the automated tool.

4   Research Method

As stated in the introduction, high-level effort estimation is a process that can be resolved as a classification problem with several imbalanced classes. Questions for determining complexity of the requested services have been designed as binary decisions on purpose. Thanks to the closed character of each question, it was easier to gather needed data, and the processing of the gathered dataset was straightforward. Also, with having only two possible answers it was easy to complete questionnaires for each of the new services. Keeping in mind the classification nature of given problems and a history of generalization complexity as a set of binary decisions it seems compelling to use the decision tree-based approach for such problems. As a next stage in achieving stable and reproducible processes for software development effort estimation it was decided to start development of random forest classifier as a classification model. During the research, authors have compared the accuracy of three different approaches to the classification problem. Those methods are respectively: decision tree classifier, extra tree classifier, and random forest classifier. Each of the classifiers has been implemented with Python and scikit-learn library. The data set used for training was a set of epics collected during ongoing digital transformation in the major investment banks. Data was collected across three major releases which contained twelve two-week sprints. The project goal was about providing digitization and automation of existing business processes within the bank, with ServiceNow as a primary automation engine. In total, 957 epics were gathered. This data set has been split into five classes: Class-0 story points in a range between 50–99 (123 epics), Class-1 story points in a range between 100–149 (387 epics), Class-2 story points in a range between 150–199 (151 epics), Class-3 story points in a range between 200–249 (117 epics) and Class-4 story points in a range between 250–300 (179 epics). Class distribution diagram is visualizing data used for the research, visualization is shown in Fig. 1. Values proportions for each of the features are presented in Fig. 2. The data set has been split into training and testing sets in nine to one ratio. For all the classifiers binary features described in 4.1. were used. Each of the models has been trained for one hundred times on the same data set and when possible, using the same hyperparameters. Training was done with the usage of the whole training set. Testing was not split into test and validation sets. One of the authors' hypothesis was that models based on implementation of random forest classifiers should have the best performance as ensemble-based models are well known to have promising outcomes [13]. Desirable target of accuracy that researchers have aimed for was set to 80% accuracy.

4.1   List of Features Used for Classifiers Training, Listed by Representation in Train/Test Sets

- Feature X1

  - Is a new form needed?
  - Possible answer: Yes/No,
  - A parameter that is indicating if a requirement relates to the creation of a new form

- Feature X2

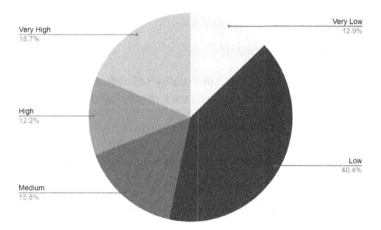

Fig. 1. Class distribution diagram

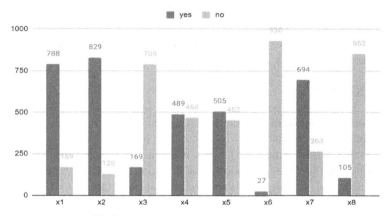

Fig. 2. Features values proportions diagram.

– Is a new field needed?
– Possible answer: Yes/No,
– This question refers to a new component for a portal form like in the past there have been created a list collector, table, solution, or any other macro or any field type which is not available currently on portal forms. Also, in the future, we might create an interactive map like a new component end to end.

• Feature X3

– Is its enhancement of existing feature?
– Possible answer: Yes/No,
– This question refers to an existing component enhancement for portal forms like in the past we have enhanced table component in case of Load Balancing Epic

- Feature X4

  - Is logic implementation needed?
  - Possible answer: Yes/No,
  - This question refers to any new functionality which is needed in portal forms, for example in the past we have implemented resubmit functionality

- Feature X5

  - Is a number of fields in the form greater than eighteen, as this is median of fields needed per new epic?
  - Possible answer: 0–18/19 +,
  - This question refers to the number of questions asked to a user on the portal form

- Feature X6

  - Is an enhancement of existing logic needed?
  - Possible answer: Yes/No,
  - This parameter is checking if more complexity is planned for the already existing solution.

- Feature X7

  - Is an implementation of new integration needed?
  - Possible answer: Yes/No,
  - Integration was identified as high risk and complex requirement. This is hard to estimate as most of the integrations need to be done with third-party vendors. With historical data, the team was able to predict that any integration will increase time to delivery

- Feature X8

  - Is the enhancement of existing integration needed?
  - Possible answer: Yes/No,
  - Any enhancement should be treated with similar complexity to new integration as it is dependent on a third party.

5   Results

During the study, research has been able to prove base hypotheses as true. As expected at the beginning of the experiment, in our research authors found that the best performing model is a random forest classifier. The training process was the longest, but the model performance was also best out of three verified models. Mean accuracy after training the model is around 78.1%. Authors were expecting a bigger difference in preference in comparison to the two other classifiers used during research which respectively achieve mean accuracy around 77, 9% for extra tree classifier and mean accuracy around 73.6% for decision tree classifier. Table 2 is containing the experiment results. After getting such promising outcomes of the experiment, the program management board has decided to give a green light for commercial solution implementation.

Table 2. Experiment results

Metric	Value
Mean accuracy for random forest classifier	0.781
Mean accuracy for extra tree classifier	0.779
Mean accuracy for decision tree classifier	0.736
Median accuracy for random forest classifier	0.787
Median accuracy for extra tree classifier	0.781
Median accuracy for decision tree classifier	0.744
Accuracy standard deviation for random forest classifier	0.033
Accuracy standard deviation for extra tree classifier	0.014
Accuracy standard deviation for decision tree classifier	0.042
Accuracy variance for random forest classifier	0.001
Accuracy variance for extra tree classifier	0.001
Accuracy variance for decision tree classifier	0.002

5.1 Key Findings

The authors have come up with several interesting findings. The researched method performed with acceptable accuracy, which was 2% below the desired target of 80%. It is also needed to mention that the feature selection process is not generic. There is a high probability that outcomes could be hard to reproduce in different projects, teams, or company setup. It is important to be aware of the fact that complexity can be affected by many independent factors. The process of selecting features should be well designed and adjusted to specific circumstances. While defining the process of software development effort estimation, it appeared requested services with complexity a level of low or very low could be treated as reference points and be included in a library of base reusable components. Thanks to this approach, during release planning the product owner was able to plan more complex features out of less collated building blocks. Another benefit of having a library of reusable reference components was to have the possibility to choose between scope and time to production delivery. As an outcome of all above average lead time for new feature delivery has shortened by 32%. As this sounds really promising, it is important that teams' velocity was almost constant and what actually have changed was that during the planning sessions it was possible to break down new feature requests into smaller work items with lower complexity.

6 Discussion

A new service has been implemented with a backend solution implementing the model. For the rest of the project, during the high-level effort estimation for the purposes of road map planning, the implemented model has been used. After the production implementation authors have registered accuracy of effort estimation around 78%. During their research authors have not focused on automated feature selection algorithms. Algorithms such as extreme gradient boosting or fuzzy trees classifiers can be incorporated for the automated feature selection process. Such an approach, together with a broader spectrum of features can lead to better accuracy and overall model performance. Automated effort estimation should not always be treated as an accurate method for creating high-level plans and project road maps. Such techniques are highly dependent on the feature selection process, size of training data set, or quality of the data itself, but at the same time tools such as automated classifiers can help with a better understanding of problems complexity. The method was introduced to help understand the complexity of the high-level requirements and to understand external dependencies. Significant positive outcome to discuss complexity and dependencies between development teams which have impact on time delivery. This leads to more accurate planning approaches and more predictable road maps. Practitioners and researchers that would like to achieve similar results should pay attention to initial feature selection, independence between estimated work items and environment setup.

7 Conclusion

This research aim is to improve on task effort estimation in agile software development processes. Findings suggest that using a random forest classifier based on yes/no questions that are tailored for the project leads to a better effort estimation in comparison to expert based estimation. After the review of the commitment for 2019 delivery, it was clear that the roadmap needs to be planned again and the scope should be adjusted to a given timeline. The high-level effort estimation tool was also having a positive impact on improved transparency across program teams. It was possible to have an estimated effort to plan more accurately services & capabilities onboarding. Going further, with the known capacity, it was easier for the Product Owner to plan specific releases and place them at the specific point of time. Authors understood that there is no silver bullet and each company has its different flavors that the software engineering industry has. Looking closely at the estimation practices used across the software industry for decades, it appears that more companies overtime is turning to modern management techniques such as agile methodologies. At the same time software development effort estimation starts to play a more and more important role in achieving desirable project outcomes. That's why it is important to have a structured and reproducible approach to this problem.

Acknowledgements. Authors are eager to share their experience regarding model implementation. Anonymized data used for training and testing the models could be shown upon request to the authors.

References

1. Braga, P.L., Oliveira, A.L.I.: Software effort estimation using machine learning techniques with robust confidence intervals. In: 19th IEEE International Conference on Tools with Artificial Intelligence (2007)
2. Patel, B.N., Prajapati, S.G., Lakhtaria, K.I.: Efficient classification of data using decision tree. Bonfring Int. J. Data Min. **2**(1), 06–12 (2012)
3. Bhatia, S., Attri, V.K.: Implementing decision tree for software development effort estimation of software project. International Journal of Innovative Research in Computer and Communication Engineering, vol. 3, no. 5, May 2015
4. Kumari, S.: Comparison and analysis of different software cost estimation methods. (IJACSA) International Journal of Advanced Computer Science and Applications, vol. 4, no. 1 (2013)
5. Geurts, P., Ernst, D., Wehenkel, L.: Extremely randomized trees. Springer Science + Business Media, 2 March 2006
6. Breiman, L.: Random forests. Mach. Learn. **45**(1), 5–32 (2001)
7. Shepperd, M., Schofield, C.: Estimating software project effort using analogies. IEEE Trans. Softw. Eng. **23**(11), 736–743 (1997). https://doi.org/10.1109/32.637387
8. Miranda, E.: Improving subjective estimates using paired comparisons. IEEE Softw. **18**(1), 87–91 (2001)
9. Vicinanaza, S., Mukhopadhyay, T., Prietula, M.: Software-effort estimation: an explaratory study of expert performance. Inform. Syst. Res. **2**(4), 243–262 (1991)
10. Boehm, B., et al.: Software Cost Estimation with COCOMO II. Prentice-Hall, Englewood Cliffs (2000)
11. Jorgensen, M.: Methods for estimating agile software projects: a systematic review. In: The 30th International Conference on Software Engineering and Knowledge Engineering (2018)
12. West, D., Kong, P., Bittner, K.: Nexus framework for scaling scrum, the: continuously delivering an integrated product with multiple scrum teams. Addison-Wesley Professional (2017)
13. Zhao, Y., Zhang, Y.: Comparison of decision tree methods for finding active objects. Adv. Space Res. **41**(12), 1955–1959 (2008)
14. Wińska, E., Dąbrowski, W.: Software development artifacts in large agile organizations: a comparison of scaling agile methods. In: Poniszewska-Marańda, A., Kryvinska, N., Jarząbek, S., Madeyski, L. (eds.) Data-Centric Business and Applications. LNDECT, vol. 40, pp. 101–116. Springer, Cham (2020). https://doi.org/10.1007/978-3-030-34706-2_6
15. Scott, E., Pfahl, D., Hebig, R., Heldal, R., Knauss, E.: Initial results of the HELENA survey conducted in estonia with comparison to results from sweden and worldwide. In: Felderer, M., Méndez Fernández, D., Turhan, B., Kalinowski, M., Sarro, F., Winkler, D. (eds.) PROFES 2017. LNCS, vol. 10611, pp. 404–412. Springer, Cham (2017). https://doi.org/10.1007/978-3-319-69926-4_29
16. Tell, P., et al.:What are hybrid development methods made of? an evidence-based characterization. In: International Conference on Software and Systems Process (2019)

MSFL: A Model for Fault Localization Using Mutation-Spectra Technique

Arpita Dutta[1] and Sangharatna Godboley[2]([⊠])

[1] Indian Institute of Technology Kharagpur, Kharagpur, India
arpitad10j@iitkgp.ac.in
[2] National Institute of Technology Warangal, Hanamkonda, India
sanghu@nitw.ac.in

Abstract. Fault localization (FL) is the most time-consuming and tedious task, while debugging. Several good techniques have been proposed for effective fault localization. These effective techniques justify the *Lean* methodology, where the waste process usually been avoided. However, most of the techniques are suffering with the problem of limited accuracy. To overcome the weakness of a technique there is a need of refinement and up-gradation of that technique. To achieve this, we can hybridize two different techniques to take advantages of both the techniques. In this paper, we propose to hybridize Mutation based testing with Spectrum based fault localization. This is a fact that both the techniques are rich in their domains. In our work, we are combining best of these techniques. We first create several mutants and drive along with the test cases to produce spectra for each mutant. This process is accountable under Agile Software Testing. These generated spectra for all mutants are supplied to fault localization techniques such as *Tarantula, Barinel, Ochiai,* and *DStar* to generate the statement ranking sequence for each mutant. Similarly, we compute the spectra for faulty program and also the statement ranking sequence. Based upon the similarity between the statement ranking sequence of faulty program and mutants, the bug is localized to most similar mutated line. We have experimented with nine open-source programs and achieved 36.48% improvement over existing FL techniques.

Keywords: Fault localization · Mutation testing · Debugging · Agile software testing

1 Introduction

Agile software testing involves all stakeholders of the team, with special expertise contributed by software testers. Software testing is one the software development phases such as requirements, design and coding. Software testing takes place simultaneously through the Software Development Life Cycle (SDLC). Agile software testing covers all the levels of testing and all types of testing. Agile

A. Przybyłek et al. (Eds.): LASD 2021, LNBIP 408, pp. 156–173, 2021.
https://doi.org/10.1007/978-3-030-67084-9_10

software testing is a software testing practice that follows the principles of agile software development.

With the increasing size and complexity of software systems, faults have become inevitable [43]. During software maintenance [26], the task of software bug localization is typically the most tedious and time-consuming [30,46]. Therefore, any improvements to the localization process can help to significantly reduce the software maintenance costs. Fault localization is also an essential part of automatic program repair (APR) techniques [26]. APR techniques rely on FL techniques to generate the search space at statement level granularity. Hence, there is a pressing need for development of an effective fault localizer. It is therefore not surprising that in the past few decades, several researchers have focused on this problem [1,3,4,12,14,19,25,30].

Weiser [1] proposed program slicing as an effective FL technique. Later, Korel et al. [2] extended Weiser's approach by considering the run-time information of test cases and named their approach as dynamic slicing. Other extensions to the slicing technique such as hybrid slicing, critical slicing, etc. [6] have also been reported. Jones et al. [12] proposed the spectrum-based fault localization (SBFL) technique. In the SBFL technique, run-time execution information of program elements (such as statements, branches, predicates, functions, etc.) is collected for different test cases along with the test execution results (success or failure). With this information, SBFL methods generate a ranked list of program elements using a mathematical formula. Tarantula [12], DStar [25], Crosstab [19], Ochiai [17], are some of the well-known SBFL techniques.

Machine learning techniques such as support vector machine [15], decision trees [13], ensemble classifiers [31] etc. have been also used to solve the problem of fault localization. Using statement coverage and test case execution information, different neural network (NN) models such as, back-propagation [14], radial basis neural network [20], deep neural network (DNN) [28], contextual information appended DNN [29], hierarchical DNN [30], convolution neural networks [32] have been used to localize the faulty statement. Though NN based techniques are popular, they require considerable training time and estimation of several parameters.

Over the last few decades, mutation testing has became very popular [33]. However, mutation testing technique has rarely been used for software fault localization. Papadakis et al. [21] reported that mutant testing is useful for fault localization as they generate appropriate substitutes for real faults. Only a few methods, such as, Metallaxis-FL [24], MUSE [23] have used mutation as a tool for locating bugs in programs.

Existing fault localization techniques make use of coverage information and test case execution results [4,12,26]. Also, the available FL techniques require examination of at least 35%–40% of program code to localize a bug [25].

In the light of these limitations, we propose a mutation testing-based fault localization technique. In this technique, we generate almost all possible mutants for a program and store them for the further use. After a new release of the system, if any test failure is reported, the similarity between faulty program spectra and test execution results with spectra and execution results of the program mutants is compared to localize the faulty statement. At present, we have considered programs with single faults only. Our proposed technique can be intuitively argued to be efficient as it only requires comparison among the raking sequences only.

In Agile process, the software is written in cycles and test cases are developed to ensure the software is correct and to protect against regression testing. However, software tests might give a false sense of security. The Mutation Testing is a technique which analyses the thoroughness of a test suite and helps identify which lines are not tested exhaustively. As we have discussed above the Mutation Testing is very costly both in terms of execution time and the time it takes a developer to analyse mutation results. In our work we tried to solve the issues and evaluate the concept of Localised Mutation (faults). Fault Localisation exploits the fact that in this modern age of agile software development, software is written in iterations. By only considering the additions or modifications to the source code, the number of mutants generated is drastically reduced. This makes Mutation Testing more feasible and in turn reduces the cost of software development as Mutation Testing can be used to detect bugs in earlier stages of development where bugs cost much less to fix. In this work we will try to answer the question, "**RQ:** Can Hybrid Mutation Testing be made efficient to be practical for everyday use, particularly for agile environments?"

Rest of the paper is organized as follows: we first review the literature of fault localization techniques in Sect. 2. In Sect. 3, we present our proposed approach. Experimental results are discussed in Sect. 4. In Sect. 5, we present the comparison of our proposed techniques with related FL approaches. Finally, we conclude this article in Sect. 6 with some future insights.

2   Related Work

Weiser [1] introduced the concept of program slicing for localizing software faults. Program slicing is based on the idea that if a test case fails to generate the correct value for a variable at a given statement. Then, the fault must exists in the static backward slice associated with the variable-statement pair. Later, Korel et al. [2] extended Weiser's approach by adding the run-time execution information of the test cases and named it *dynamic* slicing. A dynamic slice includes only those statements that are executed by the failed test case. It reduces the search domain space for localizing the fault. Another extensions for slicing such as hybrid slicing [6], thin slicing [44] etc. have been proposed.

A shortcoming of slicing based techniques is that, many times the faulty statement is not present in the slice. Also, when the faulty statement is present in the slice, too much code is required to be examined. Another limitation is that, it does not assign any ranking to the statements.

Another family of fault localization techniques is spectrum based. They are popularly known as SBFL (Spectrum Based Fault Localization) techniques. In this, coverage information of program elements such as statements, blocks, functions etc. and test case success/failure information are used as inputs. These techniques calculate the suspiciousness scores of statements using different mathematical formulas. Tarantula [12] is a well-known SBFL technique. An experimental study over Siemens suite [40] suite reported that Tarantula performs more effectively than set-union [8], cause-transition [11], set-intersection [8], and nearest-neighbor [10] methods for FL. Later on, several other SBFL techniques such as Ample [26], Crosstab [19], Barinel [42], Ochiai [17] etc. are reported. DStar(D*) [25] is a prominent SBFL technique. It is based on binary similarity coefficient and derived by *Kulczynki* coefficient. The * present in the D* is a numerical value and it varies from 2 to 50. Wong et al. [25] shows that for (*=2), D* performs better than all the existing SBFL techniques for most of the programs.

SBFL techniques are both effective and efficient. But, these techniques suffer with the problem of *ties* [18]. A tie occur when the same suspiciousness score is assigned to two or more number of statements. The probability of the assignment of same suspiciousness scores to different statements is increases with the increase of the program size.

Machine learning techniques are also used to solve the problem of fault localization. Wong et al. [14] introduced the usage of back-propagation neural network (BPNN) model for FL. But, due to the limitations like local-minima [7,27] and paralysis [5] of BPNN, Wong et al. [20] shifted to radial basis function neural network (RBFNN) model. However, both the BPNN and RBFNN models are shallow in architecture and unable to perform well with limited training data. To solve these issues, Zhang et al. [29] proposed to use deep-neural network (DNN) models for FL. Zheng et al. [28] improved the DNN based FL approach by appending contextual information into it. Dutta et al. [30] proposed a hierarchical approach for effective fault localization using DNN. They first localize the faults at the function level and next to the statement level. Zhang et al. [32] proposed to use convolution neural network (CNN) model for FL. Machine learning based FL techniques are effective but they require large time for training and parameter fixation.

Despite wide usage of the available FL techniques, these methods suffer from the problem of coincidental correctness. In coincidental correctness, statements executed by failed tests may not have caused the test to fail and also faulty statements may get executed by passed tests coincidentally [33]. To mitigate this gap, mutation-based fault localization (MBFL) techniques were introduced. Mutation testing is a popular technique for test case generation and bug prediction, but it is rarely used for fault localization [24]. It was believed that mutation is very expensive and difficult to scale. However, mutation testing has a strong capability to replicate real-world bugs. Now-a-days, several open source mutation testing tools such as MILU[1], PIT[2], Javalanche[3] are available. Also the computational capability of computers have increased phenomenally.

---

[1] https://github.com/yuejia/Milu.

[2] https://pitest.org/.

[3] http://www.javalanche.org/.

Zhang et al. [22] proposed a framework to combine the potential impacts of edits as well as the spectrum information of edits for more accurate fault localization. The potential impacts of edits are simulated by mutation changes. Whereas, the spectrum information of edits are obtained using FAULTRACKER application [16], a tool developed by the authors. Later, Moon et al. [23] introduced an FL technique entirely based on mutation testing, i.e., MUSE. The key idea of MUSE is to identify the statements in the faulty program for which the created mutant generates less number of failed test cases and more pass test cases than the actual faulty program. They have also proposed an evaluation metric for FL methods based on the concepts from information theory and named it as LIL (Locality of information loss).

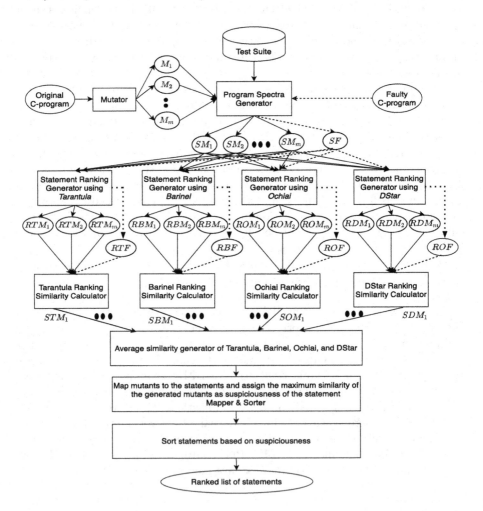

Fig. 1. Schematic representation of MSFL (Mutation-Spectra for Fault Localisation)

3 Proposed Approach

In this section, we discuss our proposed MSFL approach in detail. We named our approach as MSFL which stands for **M**utation-**S**pectrum based **F**ault **L**ocalization.

3.1 Overview

We assume that the original program has been thoroughly tested and therefore is fault-free and the test suite is available. Later, due to maintenance, bugs get introduced in the program, and we need to localize those bugs. Figure 1 presents the basic idea of our proposed MSFL using a flow chart. As it has been shown in Fig. 1, the input to the MSFL is a faulty program, and output is a ranked list of statements based on their suspiciousness of containing a fault.

First, the original program is executed with available test suite and all the outputs are saved in separate files. We create large number of mutants for the original C-program using the developed tool *mutator*. Let us consider, for a bug-free C-program P, we create m number of mutants. The ith mutant is represented as M_i. The value of i is between 1 to m. The faulty program for which we have to localize the bug is represented as F. In Fig. 1, the solid lines represent the flow of execution of the mutated versions and the dotted line shows the activities performed over the faulty program F. All the mutants along with the faulty program is supplied to the Program Spectra Generator (SPG). SPG generates the statement coverage information and test execution results for each mutant and the faulty program using available test suite. SM_i and SF represent the spectra generated for the ith mutant and the faulty program respectively.

All the generated program spectra are supplied to the Statement Ranking Generators (SRG). SRG module is loaded with a SBFL (Spectrum Based Fault Localization) technique to generate the ranking sequence of the statements. In our MSFL model, we use four SBFL techniques: Tarantula [12], Barinel [42], Ochiai [17], and DStar [25]. Tarantula [12] is a widely accepted fault localization technique and it performs better than set intersection [8], set union [8], cause-transition [11], and nearest neighbour [10] techniques of FL. Ochiai [17] is reported to be more effective than Tarantula [8,12] and many other SBFL techniques. Barinel [19] is another prominent SBFL technique. DStar [25] is considered as state-of-the-art SB-fault localization techniques.

In the Fig. 1, RTM_i, RBM_i, ROM_i, and RDM_i, represent the ranking sequence generated by Tarantula, Barinel, Ochiai, and DStar techniques for the ith mutant respectively. Similarly, RTF, RBF, ROF, and RDF, shows the ranking sequences generated for the faulty program by Tarantula, Barinel, Ochiai, and DStar methods respectively. Subsequently, the raking sequences generated for the mutants and the faulty program is supplied to the Ranking Similarity Calculator (RSC). This module computes similarity between the statement raking sequence of faulty program with statement ranking sequence of each mutant using Kendall's tau correlation coefficient [45]. Followed by this, we get four similarity score for each mutant corresponding to the four SBFL techniques undertaken. STM_i, SBM_i, SOM_i, and SDM_i, represent the similarity scores obtained by Tarantula, Barinel, Ochiai, and DStar techniques for the ith mutant respectively.

Further, the scores of generated by these four techniques are supplied to the Average Similarity Generator (ASG) module. ASG module combines these scores by taking the average of them and outputs the average similarity of each mutant with the faulty program. Based on the average similarity score, the bug is localized to the mutated statement for which the failed program has the most similar behavior.

A number of mutants may get generated for a single statement. So, the similarity score of the mutant with the highest value is assigned to the suspiciousness score of the corresponding statement. Subsequently, the statements are sorted based on their similarity scores in decreasing order and the ranked list of the statements is returned as output.

3.2   Detailed Description

Mutator. The first module used in our approach is the mutator. It takes a C-program as an input and creates different mutants for the input program. We consider only those faults for which the possible number of valid substitutes is fixed (finite). For example, valid mutants for the relational operator ('<') are only the five other operators from that group ('>', '≤', '≥', '==', '! =') without inducing any syntactic error. Table 1 shows the fault classes used for generating mutants.

Table 1. Mutation operators

Operator	Description	Example
AOR	Arithmetic operator replacement	a + b → a − b
LOR	Logical operator replacement	a || b → a && b
ROR	Relational operator replacement	a < b → a > b
CNF	Condition negation fault	a || b → !a || b
PNF	Predicate negation fault	a || b → !(a || b)

Table 2. Sample program spectra with test case execution results

Test case	S_1	S_2	S_3	S_4	S_5	S_6	S_7	S_8	S_9	S_{10}	Result
TC_1	1	0	0	0	0	0	1	1	0	0	P
TC_2	1	1	1	1	1	1	1	0	0	1	F
TC_3	1	0	0	0	0	0	1	1	0	1	F
TC_4	1	1	1	1	1	1	1	0	1	1	P
TC_5	1	0	1	0	0	0	0	1	1	1	P
TC_6	1	1	1	1	1	1	1	0	0	1	F
TC_7	1	0	1	0	0	0	0	1	1	1	P
TC_8	1	1	1	1	0	0	0	0	1	1	F

Program Spectra Generator (PSG). It takes a C-program and its test suite as input and generates the statement coverage information (spectra) and test execution result as outputs. If a statement is executed by a test case then it is represented as '1' otherwise as '0' in the spectra. The test case execution result shows whether the test case is pass or failed. If the actual output of the test case is equivalent to the expected output then the test case is considered as pass otherwise considered as failed. A pass test case is represented as 'P' and failed test case is represented as 'F'. Table 2 shows an example of program spectra with test execution result. In this example, the C-program contains ten executable statements and the test suite size is eight. Among the eight test cases, four test cases (TC_2, TC_3, TC_6, and TC_8) are failed and remaining four test cases (TC_1, TC_4, TC_5, and TC_7) are passed. It can be also observed that only the statements (S_1, S_7, and S_8) are executed by the TC_1 and remaining statements are not covered by it.

Table 3. Symbols used in the paper

N	Total number of test cases
$N_p(s)$	Total number of passed test cases
$N_f(s)$	Total number of failed test cases
$N_e(s)$	Total number of test cases executed statement s
$N_n(s)$	Total number of test cases not executed statement s
$N_{ep}(s)$	Total number of passed test cases executed statement s
$N_{ef}(s)$	Total number of failed test cases executed statement s
$N_{np}(s)$	Total number of passed test cases not invoked statement s
$N_{nf}(s)$	Total number of failed test cases not invoked statement s

Statement Ranking Generator (SRG). It takes the program spectra information along with the test execution result as an input. Then, it generates the statement ranking sequences based upon an SBFL technique as output. Table 4 shows the formulas of the four SBFL techniques used in our proposed MSFL technique. The notations used in Table 4 are defined in Table 3. The SRG component, first computes the suspiciousness scores of the statements using an SBFL technique. Subsequently, it ranks the statements based on the suspiciousness scores. Since, SBFL techniques assign same suspiciousness scores to two or more number of statements, we assign worst case rank to those statements in that situation. Table 5 shows the example statement ranking sequences generated for the five example mutants using the Tarantula SBFL technique. Similarly, SRG generates the statement rankings for the faulty program too.

Table 4. Spectrum based fault localization techniques and their formulas

S. No	SBFL technique	Formula
1	Tarantula [12]	$\dfrac{\frac{N_{ef}(s)}{N_{ef}(s)+N_{nf}(s)}}{\frac{N_{ef}(s)}{N_{ef}(s)+N_{nf}(s)}+\frac{N_{ep}(s)}{N_{ep}(s)+N_{np}(s)}}$
2	Ochiai [17]	$\dfrac{N_{ef}(s)}{\sqrt{(N_f)*(N_{ef}(s)+N_{ep}(s))}}$
3	Barinel [42]	$1-\dfrac{N_{ep}(s)}{N_{ep}(s)+N_{ef}(s)}$
4	DStar(D*) [25]	$\dfrac{(N_{ef}(s))^*}{(N_{ep}(s))*(N_{nf}(s))}$

Ranking Similarity Generator (RSG). It computes the similarity between the ranking sequence of faulty programs and mutants. It uses the Kendall Tau Rank Correlation Coefficient (τ) [45] to measure the ordinal connectivity between two ranking sequences. The value of these coefficient lies in the range of $[-1, 1]$. If any two sequences are fully correlated or similar, then the τ value is 1. On the other hand, for completely dissimilar sequences, it results in -1. The value of τ is computed using Eq. 1.

$$\tau = \frac{ct - dt}{n(n-1)} \tag{1}$$

Where, ct and dt represent the number of concordant and discordant pairs respectively and n shows the number of elements present in the ranking sequence.

Table 5. Example Statement Ranking Sequences generated by Tarantula

Mutants	RS_1	RS_2	RS_3	RS_4	RS_5	RS_6	RS_7	RS_8	RS_9	RS_{10}
M_1	10	3	8	8	1	4	5	2	6	10
M_2	10	1	3	3	4	8	5	6	7	10
M_3	10	3	5	5	1	2	7	6	5	10
M_4	10	1	4	4	2	3	6	7	8	10
M_5	10	5	7	7	2	1	3	4	8	10

Let us consider, the ranking sequence generated for the faulty program is given in Table 6. The similarity of ranking sequence of faulty program with ranking sequences of mutants is ((M1, 0.2558), (M2, 0.8606), (M3, 0.3954), (M4, 0.5349), (M5, 0.1163))

Table 6. Example Statement Ranking Sequence for faulty program

Faulty program	RS_1	RS_2	RS_3	RS_4	RS_5	RS_6	RS_7	RS_8	RS_9	RS_{10}
F	10	1	3	3	6	8	5	4	7	10

Average Similarity Generator (ASG). ASG takes the similarity scores for each mutant generated for all the four SBFL techniques as input. It combines the similarity scores of all the SBFL techniques and assign the average score of all these four techniques as to the mutant. The average similarity score of the ith mutant is computed using Eq. 2.

$$Avg\_score(M_i) = \frac{STM_i + SBM_i + SOM_i + SDM_i}{4} \tag{2}$$

Where, STM_i, SBM_i, SOM_i, and SDM_i, represent the similarity scores obtained by Tarantula, Barinel, Ochiai, and DStar techniques for the ith mutant respectively.

Mapper and Sorter. It takes the similarity scores for each mutant generated from ASG as input. More than one mutant is generated for a statement. Therefore, the mapper part first maps all the mutants to their respective statements. It assigns the maximum of the similarity score of the mutants as the suspiciousness value of the statement. For example, let $M_i, M_{i+1}, M_{i+2}, ..., M_{i+k}$ are the k mutants generated for a statement S_j. The suspiciousness value of statement S_j is calculated using Eq. 3.

$$Susp\_val(S_j) = max(Avg\_score(M_i), ..., Avg\_score(M_{i+k})) \tag{3}$$

where, $Avg\_score(M_i)$ shows the average similarity score of the ith mutant with faulty program and $Susp\_val(S_j)$ is a function to compute the suspiciousness value of statement S_j. Subsequently, Sorter sorts the statements based on their suspiciousness values and outputs a ranked list of statements.

4   Experimental Results

In this section, we first discuss the used experimental setup. Followed by this, we present the details of considered subject programs and used evaluation metric. Subsequently, we present an overview of obtained results. Last, we discuss some threats to the validity of our approach.

4.1   Setup

All the experiments are performed on a 64-bit Ubuntu 18.04.3 LTS machine with 16 GB RAM and Intel®Core^{TM} processor. All the input programs considered for our study are written in ANSI-C format. We compiled the input programs with GCC-7.4.0 compiler. Program spectra (i.e., statement coverage information) and test case execution results are collected using GCOV [39] tool. GCOV is a statement by statement profiling and code coverage analysis tool. It is utility package and freely available with GNU compiler suite. For creating mutants, we have developed a mutator and it is publicly available here [34]. To developed all the modules, python is used as a scripting language.

4.2   Data-Set Used

To evaluate the effectiveness of our proposed technique, we have experimented using three different program suites comprises of total nine programs. Five programs are belong to Siemens suite and they are downloaded from SIR repository [40]. Another four programs are from NTS benchmark suite and were downloaded from NTS repository [41]. The benchmark programs were downloaded along with their corresponding test suites and faulty versions. Table 7 presents the characteristics of all the considered programs. Program name, number of faulty versions considered, lines of code, number of functions, total executable lines of code, number of test cases, and the number of mutants generated are shown in Columns 2, 3, 4, 5, and 6 respectively.

Siemens suite [40] is considered as a benchmark for evaluating effectiveness of FL techniques [12,20,25,30]. *Tcas* and *Tot_info* are used in air traffic collision systems and information measurement machines respectively. *Schedule* and *Schedule2* are priority schedulers. Last four programs (Sl. no. 5–8) are taken from NTS repository [30,41]. *adpcm* [35] is an adaptive differential pulse code modulation program. *Merge2BST* [36] combines two binary search trees with limited extra space. *nextDate* [37] calculates the date on adding a number of days to a particular date. *quick_sort* [38] is program for sorting input numbers. *Space* program was developed at European Space Agency. It takes Array Definition Language (ADL) statements, and checks for there adherence to the ADL grammar and consistency rules [40].

Table 7. Program characteristics

S. no.	Program	No. of fty. versions	Lines of code	No. of functions	Executable LOC	No. of test cases	No. of mutants
1	Tcas	36	173	9	65	1608	216
2	Tot_info	19	406	7	122	1052	447
3	Schedule	7	412	18	152	2650	206
4	Schedule2	9	307	16	128	2710	250
5	adpcm	7	916	17	270	1600	638
6	Merge2BST	8	226	7	93	197	176
7	nextDate	14	204	6	81	378	223
8	quicksort	6	99	7	59	128	102
9	Space	38	9123	136	3656	13585	17521

4.3 Evaluation Metric

To compute the effectiveness of a fault localization technique, we use *EXAM Score* metric [10]. It shows the percentage of statements are examined to localize the faulty line in the whole program. *EXAM Score* is mathematically defined using Eq. 4.

$$EXAM\ Score = \frac{|S_{examined}|}{|S_{total}|} * 100 \tag{4}$$

Where, $S_{examined}$ and S_{total} are sets which contains the statements examined to localize the fault and statements presents in the program respectively. For a faulty program P, if the *EXAM Score* of FL technique$_1$ is lesser than FL technique$_2$, then FL technique$_1$ is more effective than FL technique$_2$.

The average improvement achieved using an FL technique, say Tech$_a$, over another FL technique, Tech$_b$, is calculated using Eq. 5.

$$IA_{a,b} = \frac{Avg.ES_b - Avg.ES_a}{Avg.ES_a} * 100 \tag{5}$$

Where, $IA_{a,b}$ shows is the improvement achieved using Tech$_a$ over Tech$_b$. $Avg.ES_a$, and $Avg.ES_b$ present the average ES obtained by Tech$_a$ and Tech$_b$ respectively. Lesser the average $EXAM\_Score$ better the technique is. For example, for a program suite P, $Avg.ES_a$ and $Avg.ES_b$ are 10 and 15 and the resultant $IA_{a,b}$ is 50%. We can say that Tech$_a$ is 50% more effective than Tech$_b$ for fault localization.

4.4 Results Obtained

We present the comparative results of MSFL with five spectrum based fault localization techniques viz., Tarantula [12], Crosstab [19], Ochiai [17], DStar [25] and Barinel [42]. We have already mentioned the importance of Tarantula [12], Ochiai [17], Barinel [42] and DStar [25] techniques. The fifth prominent technique is Crosstab [19] and is a statistical analysis based method from SBFL family.

Figures 2, 3, 4, 5, 6 and 7 present the results obtained for MSFL and the other FL techniques over different program suite using EXAM_Score metric. In each line graph, the x-axis and y-axis represent the percentage of executable statements examined and the faulty versions localized respectively. A point (x, y) in the graph shows that the y% of faulty versions are successfully localised by examining an amount of code less than or equal to x% of total program statements. Since, SBFL techniques assign same suspiciousness scores to two or more number of statements. It results in two different types of effectiveness viz., *Best* case effectiveness ad the *Worst* case effectiveness. The *best* case occurs when the faulty statement is the first to be examined among the statements with the same suspiciousness score. Similarly, the faulty statement is examined last in the *worst* case. Two different plots have been used to represent the *Best case* and the *Worst case* effectiveness, DStar [25], Tarantula [12], Ochiai [17], Barinel [42] and Crosstab [19] fault localization techniques. A single line graph will be used to present the MSFL technique.

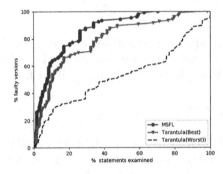 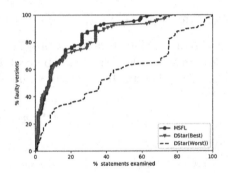

Fig. 2. Effectiveness comparison of MSFL against Tarantula

Fig. 3. Effectiveness comparison of MSFL against DStar

Figure 2 shows the effectiveness comparison of our proposed MSFL approach with Tarantula over the selected set of Siemens and NTS suite programs. It can be observed from the figure that by examining less than 2% of code, MSFL technique localizes bugs in 26.42% of faulty versions. On the other hand, Tarantula (Best) and Tarantula (Worst) localize bugs in only 15.09% and 2.83% of faulty versions by examining the same amount of program code. On an average, MSFL technique is 35.63% and 61.49% more effective than Tarantula (Best) and Tarantula (Worst) respectively.

Figure 3 represents the effectiveness comparison of DStar and MSFL techniques for Siemens and NTS suites. It shows that 75% of faulty versions are localized by examining only 20% of program code by MSFL. Whereas, for the localization of faults in same percentage of faulty versions, DStar (Best) and DStar (Worst) require to examine at least 28.81% and 75.27% of program code respectively. In the worst case, MSFL technique requires 8.51% and 29.69% less code examination than DStar (Best) and DStar (Worst). On an average, MSFL is 15.44% and 47.46% more effective than DStar (Best) and DStar (Worst) respectively.

Figure 4 shows the comparison of MSFL and Ochiai techniques over Siemens and NTS suites. It can be observed from the figure that by examining only 20% of code MSFL localizes bugs in 79% of faulty versions whereas, Ochiai (Best) and Ochiai (Worst) localize bugs in only 71% and 34% of faulty versions. In the worst case, MSFL requires 12.72% and 29.69% of less code examination than Ochiai (Best) and Ochiai (Worst) respectively. On an average, MSFL is 36.31% more effective than Ochiai. Figure 5 shows the effectiveness comparison of our MSFL technique with Crosstab for fault localization for Siemens and NTS suites. It can be observed from the figure that 35% of faulty versions are localized by examining only 4% of program code by MSFL. Whereas, for the localization of faults in same percentage of faulty versions, Crosstab (Best) and Crosstab (Worst) respectively require to examine at least 28.81% and 75.27% of program code. In the worst case, MSFL technique requires 29.69% less code examination than both Crosstab (Best) and Crosstab (Worst). On an average, MSFL is 26.74% and 58.74% more effective than Crosstab (Best) and Crosstab (Worst) respectively.

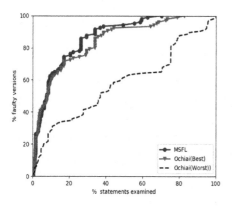

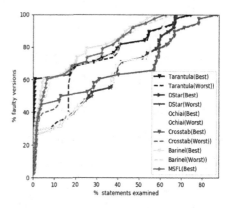

Fig. 4. Effectiveness comparison of MSFL against Ochiai

Fig. 5. Effectiveness comparison of MSFL against Crosstab

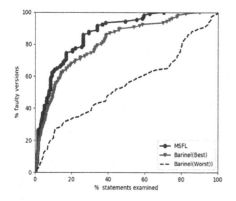

Fig. 6. Effectiveness comparison of MSFL against Barinel

Fig. 7. Effectiveness comparison of MSFL with different FL techniques over Space

Figure 6 show the effectiveness comparison of MSFL with Barinel for Siemens and NTS suite. It can be observed from the figure that by examining 10% of faulty versions only MSFL localizes bugs in 63.21% faulty versions. On the other hand for the same code examination, Barinel (Best) localizes bugs in only 50% of versions. In the worst case, MSFL technique requires 19.67% and 29.69% less code examination than Barinel (Best) and Barinel (Worst) respectively. On an average, MSFL is 50.04% better than Barinel.

Figure 7 shows the effectiveness comparison result of MSFL with DStar, Tarantula, Ochiai, Crosstab, and Barinel over the Space suite programs. It can be observed from the Fig. 7 that MSFL performs better than Tarantula, Ochiai, DStar and Crosstab for almost all the faulty versions. Only for a few versions, MSFL is less effective than Berinal. Average Exam_Score of Tarantula, DStar, Ochiai, Crosstab, Barinel and MSFL are 20.42%, 30.03%, 28.83%, 30.15%, 15.94%, and 12.46%. In the worst case, MSFL is 12.23%, 5.54%, 2.75%,

25.59%, and 4.70% better than Tarantula, DStar, Ochiai, Crosstab, and Barinel respectively.

4.5 Threats to the Validity

- At present, we have considered only a limited set of programs to evaluate the effectiveness of proposed approach. It is possible that our proposed technique may not work effectively for other set of programs. However, to mitigate this risk, we have considered programs with different size, application domain and complexity.
- The effectiveness of our approach is low if no mutants available for the faulty line. In this case, a lesser rank is assigned to the faulty compared to the non-faulty statements for which mutants are already available.

5 Discussion

Cleve et al. [11] proposed a state-model based FL technique known as cause-transition. They compared the states of different passed and failed runs to locate the cause of failure. This method is an extension of their previous work proposing *delta debugging* [9]. Jones et al. [8] reported that Tarantula [12] requires less number of statements to be examined for localizing faults compared to the cause-transitions [11], set-union [8], nearest-neighbor [10] and set-intersection [8]. From our experimental results, it can be observed that our proposed approach examines, on an average, 48.56% less statements than Tarantula [12].

Renieris et al. [10] proposed the nearest-neighbor approach for FL. They targeted to find the most similar trace generated from the successful test cases with a failed test case trace. Further, they applied a set difference to eliminate the irrelevant statements from the failed test case trace and returns a list of suspicious statements. The effectiveness of their approach is completely dependent on the used test suite. Also, in some cases, it returns a null set of suspected statements. Whereas, our proposed approach generates a ranked list of statements based on their suspiciousness of containing a fault.

A number of slicing based techniques have been reported in the literature for FL and have became popular [1,4,6]. The main drawback of this approach is sometimes it returns the complete program as a slice and thereby nullifying the effectiveness of the approach. Also, these techniques do not assign any ranks to the statements. On the other hand, our proposed approach provides a rank to each executable statement for which mutants are generated.

We have compared our proposed work MSFL with four prominent SBFL techniques: DStar [25], Tarantula[12], Ochiai [17], and Crosstab [19]. Our results indicate that our proposed approach performs 31.45%, 48.56%, 36.31%, and 42.46% better than the respective techniques. DStar is the state-of-the-art for fault localization techniques based on the program spectrum information. However, DStar excludes the information of successful test cases which do not cover the statement while calculating the suspiciousness score. On the other hand, our proposed approach took the complete information of both pass and failed test cases to calculate the suspiciousness of a statement. It helps our approach to distinguish the behavior of the statement to localize the faults.

Wong et al. [14] was the first to use neural network models for FL. First, they used BPNN [14] and later RBFNN [20] for the same. Zheng et al. [28] used a deep neural network, and Zhang et al. [29] extended [28] approach by adding the contextual information to localize the faults. Neural networks can model extremely complex functions, but there are problems of non-deterministic parameter estimation and feedback loop. Neural network models require a long time for training and parameter fixing (of the order of tens of minutes). On the other hand, the time required in each step of our proposed approach is order of seconds, even for a large-sized program.

6 Conclusion

We have presented a hybrid technique of mutation testing and spectrum-based fault localization. The proposed MSFL technique generates several mutants and compares the program spectra and test execution results to locate suspicious statements by measuring the distance with four different spectrum-based FL techniques. We have compared the effectiveness of our proposed technique with SBFL methods over nine different open-source programs. On an average, our proposed technique MSFL is 36.48% more effective than existing fault localization techniques. Finally, we justify our RQ, and we have showed that Hybrid Mutation Testing is efficient to be practical for everyday use like agile process.

At present, we have considered programs with single faults only. We plan to extend our proposed approach by mutating multiple statements at the same time and localize the multiple-fault programs. Also, we will considered machine learning approaches in addition to the SBFL techniques for generating different statement ranking sequences. It will help to improve the effectiveness of fault localization.

References

1. Weiser, M.: Program slicing. IEEE Trans. Softw. Eng. **10**(4), 352–357 (1984). https://doi.org/10.1109/TSE.1984.5010248
2. Korel, B., Laski, J.: Dynamic program slicing. Inf. Process. Lett. **29**(3), 155–163 (1988)
3. Agrawal, H., et al.: Design of mutant operators for the C programming language. Technical report SERC-TR-41-P, Software Engineering Research Center, Purdue University (1989)
4. Agrawal, H., Horgan, J.R.: Dynamic program slicing. In: Proceedings of the ACM SIGPLAN'90 Conference on Programming Language Design and Implementation, pp. 246–256. White Plains, New York (1990)
5. Wasserman, P.D.: Advanced Methods in Neural Computing. Wiley, Hoboken (1993)

6. Gupta, R., Soffa, M.L.: Hybrid slicing: an approach for refining static slices using dynamic information. In: Symposium on Foundations of Software Engineering, pp. 29–40 (1995)
7. Mitchell, T.M.: Machine Learning. McGraw-Hill, New York (1997)
8. Jones, J.A., Harrold, M.J., Stasko, J.: Visualization for fault localization. In: Proceedings of ICSE 2001 Workshop on Software Visualization, Ontario, BC, Canada, May 2001, pp. 71–75 (2001)
9. Zeller, A., Hildebrandt, R.: Simplifying and isolating failure-inducing input. IEEE Trans. Softw. Eng. 28(2), 183–200 (2002)
10. Renieris, M., Reiss, S.P.: Fault localization with nearest neighbor queries. In: Proceedings of the 18th IEEE International Conference on Automated Software Engineering, Montreal, Canada, October 2003, pp. 30–39 (2003)
11. Cleve, H., Zeller, A.: Locating causes of program failures. In: Proceedings of the 27th International Conference on Software Engineering, St. Louis, Missouri, USA, May 2005, pp. 342–351 (2005)
12. Jones, J.A., Harrold, M.J.: Empirical evaluation of the tarantula automatic fault localization technique. In: Proceedings of the 20th IEEE/ACM Conference on Automated Software Engineering, Long Beach, California, USA, pp. 273–282 (2005). https://doi.org/10.1145/1101908.1101949
13. Briand, L.C., Labiche, Y., Liu, X.: Using machine learning to support debugging with Tarantula. In: The 18th IEEE International Symposium on Software Reliability (ISSRE 2007), pp. 137–146. IEEE (2007)
14. Wong, W.E., Qi, Y.: BP neural network-based effective fault localization. Int. J. Softw. Eng. Knowl. Eng. 19(4), 573–597 (2009). https://doi.org/10.1142/S021819400900426X
15. Ascari, L.C., Araki, L.Y., Pozo, A.R., Vergilio, S.R.: Exploring machine learning techniques for fault localization. In: 10th Latin American Test Workshop, pp. 1–6. IEEE (2009)
16. Zhang, L., Kim, M., Khurshid, S.: Localizing failure-inducing program edits based on spectrum information. In: 27th IEEE International Conference on Software Maintenance (ICSM), pp. 23–32. IEEE (2011)
17. Naish, L., Jie, L.J., Kotagiri, R.: A model for spectra-based software diagnosis. ACM Trans. Softw. Eng. Methodol. (TOSEM) 20(3), 1–32 (2011)
18. Xu, X., Debroy, V., Wong, W.E., Guo, D.: Ties within fault localization rankings: exposing and addressing the problem. ACM Trans. Softw. Eng. Methodol. (TOSEM) 21(06), 803–827 (2011)
19. Wong, W.E., Debroy, V., Xu, D.: Towards better fault localization: a crosstab-based statistical approach. IEEE Trans. Syst. Man Cybern. Part C (Appl. Rev.) 42(3), 378–396 (2011)
20. Wong, W.E., Debroy, V., Golden, R., Xu, X., Thuraisingham, B.: Effective software fault localization using an RBF neural network. IEEE Trans. Reliab. 61(1), 149–169 (2012). https://doi.org/10.1109/TR.2011.2172031
21. Papadakis, M., Traon, Y.L.: Using mutants to locate "unknown" faults. In: 5th International Conference on Software Testing, Verification and Validation, pp. 691–700. IEEE (2012)
22. Zhang, L., Zhang, L., Khurshid, S.: Injecting mechanical faults to localize developer faults for evolving software. ACM SIGPLAN Not. 48(10), 765–784 (2013)
23. Moon, S., Kim, Y., Kim, M., Yoo, S.: Ask the mutants: mutating faulty programs for fault localization. In: 2014 IEEE 7th International Conference on Software Testing, Verification and Validation, pp. 153–162. IEEE (2014)

24. Papadakis, M., Traon, Y.L.: Metallaxis-FL: mutation-based fault localization. Softw. Test. Verif. Reliability. **25**(5–7), 605–628 (2015)
25. Wong, W.E., Debroy, V., Gao, R., Li, Y.: The DStar method for effective software fault localization. IEEE Trans. Reliab. **63**(1), 290–308 (2016). https://doi.org/10.1109/TR.2013.2285319
26. Wong, W.E., Gao, R., Li, Y., Abreu, R., Wotawa, F.: A survey on software fault localization. IEEE Trans. Softw. Eng. **42**(8), 707–740 (2016). https://doi.org/10.1109/TSE.2016.2521368
27. Tan, P.N., Steinbach, M., Kumar, V.: Introduction to Data Mining. Pearson Education India (2016)
28. Zheng, W., Hu, D., Wang, J.: Fault localization analysis based on deep neural network. Math. Probl. Eng. (2016). https://doi.org/10.1155/2016/1820454
29. Zhang, Z., Yan, L., Qingping, T., Xiaoguang, M., Ping, Z., Xi, C.: Deep learning-based fault localization with contextual information. IEICE Trans. Inf. Syst. **100**(12), 3027–3031 (2017)
30. Dutta, A., Manral, R., Mitra P., Mall, R.: Hierarchically localizing software faults using DNN. IEEE Trans. Reliab. (2019). (Early Access)
31. Dutta A., Pant N., Mitra P., Mall R.: Effective fault localization using an ensemble classifier. In: International Conference on Quality, Reliability, Risk, Maintenance, and Safety Engineering: QR2MSE-2019. Zhangjiajie, Hunan, China (2019)
32. Zhang, Z., Lei, Y., Mao, X., Li, P.: CNN-FL: an effective approach for localizing faults using convolutional neural networks. In: 2019 IEEE 26th International Conference on Software Analysis, Evolution and Reengineering (SANER), China, pp. 445–455. IEEE (2019)
33. Li, X., Li, W., Zhang, Y., Zhang, L.: DeepFL: integrating multiple fault diagnosis dimensions for deep fault localization. In: Proceedings of the 28th ACM SIGSOFT International Symposium on Software Testing and Analysis, pp. 169–180 (2019)
34. https://github.com/ArpitaDutta/Mutator_
35. http://www.mrtc.mdh.se/projects/wcet/wcet_bench/adpcm/
36. https://www.geeksforgeeks.org/merge-two-bsts-with-limited-extra-space/
37. https://www.geeksforgeeks.org/date-after-adding-given-number-of-days-to-the-given-date/
38. https://www.tutorialspoint.com/data_structures_algorithms/quick_sort_program_in_c.htm
39. http://man7.org/linux/man-pages/man1/gcov-tool.1.html
40. https://sir.csc.ncsu.edu/portal/index.php
41. https://github.com/ArpitaDutta/NTS_Repository
42. Abreu, R., Zoeteweij, P., Van Gemund, A.J.: Spectrum-based multiple fault localization. In: 2009 IEEE/ACM ICSE, pp. 88–99. IEEE, November 2009
43. Dutta, A., Kumar, S., Godboley, S.: Enhancing test cases generated by concolic testing. In: Proceedings of the 12th Innovations on Software Engineering Conference, pp. 1–11, February 2019
44. Sridharan, M., Fink, S.J., Bodik, R.: Thin slicing. In: Proceedings of the 28th ACM SIGPLAN Conference on Programming Language Design and Implementation, pp. 112–122, June 2007
45. Knight, W.R.: A computer method for calculating Kendall's tau with ungrouped data. J. Am. Stat. Assoc. **61**(314), 436–439 (1966)
46. Cui, Z., Jia, M., Chen, X., Zheng, L., Liu, X.: Improving software fault localization by combining spectrum and mutation. IEEE Access **8**, 172296–172307 (2020)

Short Papers

Implementing Lean Principles in Scrum to Adapt to Remote Work in a Covid-19 Impacted Software Team

Leigh Griffin[✉]

Waterford Institute of Technology, Waterford, Ireland
20006077@mail.wit.ie
http://www.wit.ie

Abstract. Agile Frameworks are a mainstay of Software Engineering teams, with Scrum having risen to prominence over the last decade. Teams are experimenting more with variants of Scrum, with inspirations coming largely from other process improvement methodologies. However, the prevalent mode of executing Scrum assumes that the team are in colocation. In early 2020, a global pandemic has shifted the world unexpectedly into remote work. The heavyweight nature of Scrum is leading to fatigue within teams, as the remote nature of video conferencing and asynchronous communication bring an overhead not experienced before. The result is a weakening of the Scrum fundamentals in teams, with modifications being made to accommodate the new normal that teams are experiencing. Not all of the changes are positive and a lot of waste has emerged within teams. Lean principles can be applied to Scrum with minor adjustments and minimal friction. The end result is a variant of Scrum designed with remote teams in mind. This paper will explore some of the challenges remote teams are facing, the modifications proposed and the result of those changes in a remote Software Engineering Team.

Keywords: Covid-19 · Lean · Remote teams · Scrum

1 Introduction

Software Development, compared to more established industries, is still in it's infancy. The frameworks to guide the creation of a software product are maturing rapidly, with inspirations drawn from other industries shaping and evolving the models over time. Scrum, has grown in popularity in recent years and a trend has emerged to move away from a pure Scrum approach, Robinson and Beecham (2019), which is Scrum as per the Scrum Guide (Schwaber and Sutherland (2018)), to move towards a hybrid model. The hybrid aspect often involves modifying some of the core ceremonies or bringing in additional changes, often inspired by other frameworks. One of the most popular hybrid approaches is the introduction of Kanban. The term Scrumban, as mentioned in Nikitina et al. (2012) has emerged to describe this model with some of the key benefits being

© Springer Nature Switzerland AG 2021
A. Przybyłek et al. (Eds.): LASD 2021, LNBIP 408, pp. 177–184, 2021.
https://doi.org/10.1007/978-3-030-67084-9_11

increased visualization of work and a focus on Work in Progress (WiP) limits. This has the affect of pulling work through the system at a faster flow rate, by focusing teams on a finite number of items. The Agile Manifesto (Beck (2001)) talks about people and interactions over processes and tools. Scrum advocates strongly for in person execution for the most optimal results and control for the team, with a strong motivator being the presence of the customer and team in colocation. The enhancements or modifications to Scrum, particularly those inspired by Kanban, focus on a visual way of implementing and running the team, lending itself to in person interactions. Minimal research, such as Faniran et al. (2017), Smits and Pshigoda (2007) and Paasivaara et al. (2012) have investigated remote friendly Scrum, but no standard tools and techniques have emerged for teams to follow. In addition, the unprecedented nature of this unexpected and first global pandemic in the history of the digital era, means that the world is learning how to adapt.

Several online tools are emerging that are proving to be more than adequate to replicate in person interactions, however, teams have not fully embraced or invested in them. In early 2020, a global pandemic named covid-19, caused a paradigm shift for the worlds workforce. With global health advice advising against in person business, most industries moved into a remote way of working. A subsequent scramble to form an IT strategy that was conducive to sustained performance was now needed. This brought several challenges for teams who had a heavy focus on in person interactions and this shift to remote work came with the expectation that performance and delivering the same output would be maintained. This paper will focus on the lessons learned from a software engineering team that had a mix of remote and in office team members before covid-19 where a culture of remote first work was already prevalent. This team had been working in an Agile manner for almost 18 months, were trying to follow the Scrum Framework and had a full time Product Owner and ScrumMaster. The team executed Quarterly Planning (QP), a process whereby stakeholders would help prioritize multiple sprints worth of work to execute a larger piece of functionality in order to align across the enterprise. The team as a whole numbered 26 people divided into 4 delivery teams, each responsible for delivering standalone functionality. The work from home enforcement of family members and house mates created an environment that put strain on our team. Of particular focus was the waste generated by following the Scrum framework and this paper will focus on the key waste elements identified in both the ceremonies and the key inputs to the team, with recommended adaption to minimize waste provided.

2   Remote Scrum Challenges

A number of challenges have emerged for remote teams running Scrum in the current environment. Scrum operates on a predicable multi week cadence, broken down into ceremonies that are designed to continiously feed a team relevant work, be highly collaborative with the customer in mind and ultimately to deliver incremental value.

2.1   Scrum Ceremonies

Daily Scrums are timeboxed at 15 min. In person interactions can allow for better facilitation and control of the communication and to ensure that the timebox is adhered to. Using online video conferencing tools introduces technology as a factor, chiefly audio/visual and network issues making the tighter timebox difficult to stay within or compromises the quality of the discussion within the team

Sprint Planning is timeboxed at a maximum of 8 h for a 4 week Sprint, with that ratio of hours to sprint duration scaling appropriately. While held in person, engagement can remain high for a longer period of time than working from home can achieve. Best practices for ergonomics, as seen in Bontrup et al. (2019) states a minimum of 5 min away from the screen every hour. This can be highly disruptive to the flow of a long running meeting like Sprint Planning, where a full Agenda cannot be set ahead of time. The natural breaks in person are more difficult to translate to that ergonomic recommendation, with meetings often running to close out a section of the planning successfully. The Sprint Review suffers from a similar timeboxing issue as it recommends a 4 h meeting for a 4 week Sprint, this, however, is less of a concern in shorter sprint durations. Sprint Retrospective is one of the most valuable tools for a Scrum team to inspect and adapt. Psychological Safety is one of the hallmarks of a strong Agile team (Edmondson (2018)), that is in part detected, and acted upon by the ScrumMaster. In person, the use of body language is a key tool to help discover and assess psychological safety. In a remote world, where web-cameras may not be available for everyone, it is impossible to gain that sense of how safe the team are. This can have a major impact on the quality of the contributions and hampers the retrospective longer term.

2.2   Distractions

While participating in any of the Scrum Ceremonies, having the team present in a mental capacity is hugely important. That diminishes quickly in a remote environment, as team members are on their laptops with a plethora of noisy distractions from emails, to instant messages through to web surfing. The result is a lack of focus within the team, and while the ceremonies are executed as close to the Scrum Framework as possible, the result is a mechanical version of Scrum, rather than the Agile mindset which results in a longer term transformation. While the impact here is hard to quantify, 6 months into the pandemic, the longer term impact will be a follow on to this research paper.

2.3   Fatigue

In person interactions allow for a lot of time away from a desk and by extension from the work at hand. A trend in recent years in the IT industry is to have an office with collaboration focused areas, wellness rooms, games rooms and multiple social interaction touchpoints in common spaces like the canteen. This has a noticeable impact on human collaboration as noted by Bernstein and Turban (2018) and the passive result is a workplace that has a social element to it, which increases staff happiness. This is an important element of a team and company culture, which results in the teams weekly contribution to their project, versus their contractual obligation, being reduced. That reduction is often passively factored into project and release plans. In a work from home environment, personal and professional spaces may collide, with variable working hours. Another factor is the narrowing of the social interactions to just that of your immediate team and extended functions critical to your day job. This focused social group, coupled with the work day being closer to the contracted hours, has a magnified intensity to the work day that workers have never experienced before. Pairing this with meeting fatigue and burnout is a real possibility within the team. That can impact the overall team and the product may suffer.

3   Waste Within Scrum: Key Inputs

A focal point here is on observations about waste, or muda, in the Lean terminology, that our team have identified for later actions.

3.1   Requirements

Working with Stakeholders is a core facet of the Product Owner role, often bringing in key technical members to fore-run on feature requests to stock the Backlog. Good Product Owners often work several Sprints ahead of the development team to ensure that Sprint Planning is a smooth and efficient process (Sverrisdottir et al. (2014)). The Quarterly Planning (QP) approach used by our team helps to indicate the delivery goals for customers. QP is a maturity indication for Agile teams, where the teams velocity, their sizing analysis and overall execution are capable of predicting 12 weeks (multiple sprints) with a high degree of confidence. With this in mind, QP brings stakeholders to analyse the User Stories scoped at the right size to be delivered in that timebox. Mature products often have longer term functionality that needs to be built upon quarter after quarter, to achieve the ultimate end goal. An element of waste here is the involvement of stakeholders in the lead up to, and the execution of QP, when the teams capacity might be a fraction of their true potential due to longer running initiatives. In our team, this has been the case for 3 Quarters running due to long term initiatives that have an overarching need to deliver a complex system.

   Roadmaps are a very popular sales tool to entice Customers into longer term investment in a product offering as described in Munch et al. (2019). That involves future projection to showcase longer term direction of the Product. This involves engaging with Customers, who are attracted to some of the future roadmap, with the view to try and accelerate features by building out requirements. Roadmaps, however, are often a marketing mechanism, talking about longer term future branch points that may not align to the Product Vision or the direction the company are trying to take. This involves the Product Owner and key team members engaging on in depth requirements gathering for features that might never be realised for a variety of reasons from technical, to product positioning through to investment required in skills, people and infrastructure.

3.2 Backlogs

With the default separation of a Sprint and Product Backlog, Scrum has a funneled approach to take items into a Sprint and allow the team break them out in more detail for consumption. The granularity of the tasks matches the Definition of Done, ensuring that multiple, often very similar in nature pieces of work being identified. While the need for Quality is undeniable in software, the process for the team to take time to create tickets (a generic term for work items to progress) to capture the items needed to complete a User Story to done, is an exercise in repetition. The risk of not doing this level of breakdown is the team missing some of their Definition of Done criteria when it comes to Sprint Review.

Multiple teams often work together on a product. Mature, complex products are often viewed as sub-products, as the functionality has grown to a level of complexity that a dedicated team owns the feature set. This creates siloed backlogs, with feature teams drawing select User Stories from the Product Backlog to stock their sub-team backlog. This approach sees the members of that team attend wider Backlog refinement, and while the knowledge sharing can be invaluable, it is viewed as muda once their subsection User Stories are discussed and agreed upon. Layers of complexity build on top of this to ensure that cross sub-team coordination can exist, with Scaled Scrum Frameworks emerging which bring multiple additional meetings and coordination in an attempt to get back to a singular product view of the world. The larger the organisation, the larger the waste footprint becomes.

4   Adaptions to the Scrum Process

While the onset of the pandemic forced many of the adaptations to accelerate, the team had already worked on modifications to the Scrum framework to become a more hybrid Scrum team.

4.1 Scrum Ceremonies

Small changes remove a lot of waste and free up precious calendar time for the teams by defaulting some of the updates to an asynchronous manner. The Daily Scrum, as an example, can be delivered over instant messaging applications for consumption by the team at their own pace. Blockers and escalations can still be elevated to calls, however, the majority of Daily Scrum conversations are coordination of work. While the Scrum Guide emphasizes a focal point on Blockers, teams use the call to align on the goals for the day and recap the prior days work, this is a relic of the original intent of the Daily Scrum that prevails as it is attractive to teams for coordination benefits. Sprint Reviews often come with demos, which can be time consuming to setup and establish. Pre-recording demos and making them available to the team to consume in their own time is another key removal of waste, allowing the team focus a conversation on the demo over other communication channels.

4.2   Backlog Adaption

Backlog refinement looks at the longer term horizon for the team from a product perspective, ordering the backlog and allowing for Sprint Planning to proceed with the certainty that the highest value items will be progressed. At Sprint Planning time, the breakdown of those tickets selected into smaller, consumable pieces for the team is a necessary step. Typically, the entire team gathers for this process to benefit from the knowledge sharing, however, a percentage of a cross functional team will have no input or real value taken from the discussion on Stories outside of their immediate skillset. While the more experienced developer leads the conversation, most of the team enter a silent observation mode. Discussing the tickets in a just in time format is a heavy team burden, as the technical depth of conversations required to articulate the needs is time consuming. The adjustment proposed was to merge the backlogs from a granularity perspective, which some Scrum variants recommend Gancarczyk and Griffin (2019). The team have the certainty of the next several sprints worth of work and this allows the technical leads to investigate the User Stories as individuals, documenting their observations, their recommended implementation details and finally sharing this with the team. Over the duration of a Sprint, the team established 1 h Backlog Refinement calls, to go in and discuss at a high level the implementation detail as documented ahead of time, to answer any questions that could not have been resolved on the tickets and to finalize their plan. Strict timeboxing ensures that the team do not feel burned out and the frequent cadence of the meetings allows team members to dip in and out as necessary, with the option of watching the recording of the meeting and following the detailed breakdown in the ticket and raise questions there. The benefits here have helped streamline the QP process as a more in depth backlog view is possible, allowing for more informed decisions on what to prioritise and when and no sense of loss was reported by the move away from a Product Backlog view.

5   Process and Tools

The Agile Manifesto clearly states individuals and interactions over processes and tools. As remote teams, tooling is important and issue trackers provide a number of ease of life improvements with APIs and plugins capable of eliminating a lot of waste. Our team implemented their Definition of Done (as defined in Schwaber and Sutherland (2018)) in an automated manner by auto creating tickets for documentation, testing validation and release readiness as part of the new ticket creation. This, when combined with static code analysis tools that can provide a gating mechanism before code merge, ensures lower technical debt all around and that the DoD is adhered to. The waste removal here is significant in backlog refinement sessions and a far safer method from a quality perspective.

Software teams spend a lot of time debugging and tracking down issues and bugs reported by their users. The ability to observe the system and quickly understand where a fault might reside represents a saving in debugging time, context switching and a quicker resolution, which ultimately allows the developer to get back to more value add work faster. The investment by our team in Observability and Monitoring stacks has provided a reassuring level of Quality to complement already expansive test coverage and eliminated a large amount of muda that the team were experiencing.

The Product Owner introduced the concept of an Epic Brief – a one page summary proposal – for stakeholders and team members to simplify the process of funneling work into the team. Previously, stakeholders would have engaged with multiple groups as they evolved their idea, resulting in a lot of wasteful meetings and often the end result was discovering that the work was not aligned with the teams mission statement or simply beyond the scope of what they could deliver. This minimal required information allows the Product Owner engage more efficiently. The benefit of this approach is a fail fast approach, to discover that the work proposed was not a good fit for the team for a number of reasons. Combining this with QP, the Product Owner is able to ascertain the workload on the team as well as the size and scope of incoming requests. This helps level set expectations before too much time and effort is put into an Epic Brief that might be important to that particular stakeholder, but that will ultimately be a lower priority for the team as a whole and several quarters away based on current trending velocity and in flight needs.

6   Conclusion and Future Work

The paradigm shift to work from home is putting a strain on teams like no other experience in their lifetime. The Scrum Framework has multiple prescribed routines that are optimized for in person interactions and that when followed to the spirit of the Scrum Guide, result in the generation of waste. Having recognized this, our team has taken the first steps in eliminating this waste and moving towards a more Lean inspired version of Scrum. As this global situation is still evolving, the work from home approach looks set to remain the status quo for years to come. More work is planned by our team on investigating the tooling from a production line viewpoint, where requirements and code eventually get baked into a workable product. Deeper statistics are being gathered to make informed changes, with the more obvious waste, as documented in this paper, eliminated already. The key constituent roles of Product Owner and ScrumMaster also generate a lot of waste within a Scrum team and is a planned future research topic.

References

Beck, K.: Manifesto for agile software development (2001). http://www.agilemanifesto. org/

Bernstein, E., Turban, S.: The impact of the 'open' workspace on human collaboration. Philos. Trans. R. Soc. B Biol. Sci. **373**, 20170239 (2018)

Bontrup, C., et al.: Low back pain and its relationship with sitting behaviour among sedentary office workers. Appl. Ergon. **81**, 102894 (2019)

Edmondson, A.C.: The Fearless Organization: Creating Psychological Safety in the Workplace for Learning, Innovation, and Growth. John Wiley & Sons, New Jersey (2018)

Faniran, V.T., Badru, A., Ajayi, N.: Adopting scrum as an agile approach in distributed software development: a review of literature. In: 2017 1st International Conference on Next Generation Computing Applications (NextComp), pp. 36–40 (2017)

Gancarczyk, A., Griffin, L.: Small scale scrum. Agile Alliance Experience Reports (2019)

Munch, J., Trieflinger, S., Lang, D.: Product roadmap - from vision to reality: a systematic literature review (2019)

Nikitina, N., Kajko-Mattsson, M., Stråle, M.: From scrum to scrumban: a case study of a process transition. In: Proceedings of the International Conference on Software and System Process, ICSSP '12, IEEE Press, pp. 140–149 (2012)

Paasivaara, M., Lassenius, C., Heikkilä, V.T.: Inter-team coordination in large-scale globally distributed scrum: do scrum-of-scrums really work?. In: Proceedings of the 2012 ACM-IEEE International Symposium on Empirical Software Engineering and Measurement, pp. 235–238 (2012)

Robinson, P.T., Beecham, S.: Twins: this workflow is not scrum: agile process adaptation for open source software projects. In: Proceedings of the International Conference on Software and System Processes, ICSSP '19, IEEE Press, pp. 24–33 (2019)

Schwaber, K., Sutherland, J.: The scrum guide (2018). https://www.scrumguides.org/

Smits, H., Pshigoda, G.: Implementing scrum in a distributed software development organization. In: Agile 2007 (AGILE 2007), pp. 371–375 (2007)

Sverrisdottir, H., Ingason, H., Jónasson, H.: The role of the product owner in scrum-comparison between theory and practices. Procedia - Soc. Behav. Sci. **119**, 257–267 (2014)

Business-Oriented Approach to Requirements Elicitation in a Scrum Project

Michał Sosnowski[1], Michał Bereza[1], and Yen Ying Ng[2]([⊠]) [iD]

[1] Faculty of Electronics, Telecommunications and Informatics,
Gdansk University of Technology, Gdansk, Poland
sosna321@gmail.com, michalber1998@gmail.com
[2] Department of English Studies, Nicolaus Copernicus University, Torun, Poland
nyysang@gmail.com

Abstract. As agile methods allow requirements to emerge throughout the development process, requirements engineering activities span the entire life cycle of a system. However, requirements elicitation is poorly performed in agile teams. Instead, agile teams rely fundamentally on the use of validation. In this paper, we demonstrate that on the contrary to what agilists say, creating a requirements document up front may not necessarily be a waste of time and can be done in a light way. We report on a Scrum project in which we adopted the Business-Oriented approach to Requirements Elicitation (BORE) proposed by Przybyłek [12]. In this approach, system requirements are derived from business process models. Accordingly, the system requirements meet real business needs and there are no superfluous requirements.

Keywords: Requirements engineering · Business process modelling · Use cases · Agile

1 Introduction

Agile methods emerged as a reaction to traditional ways of software development and acknowledged that customers are unable to definitively express their needs up front [9, 11, 17]. Accordingly, in agile software development, requirements evolve through the project lifetime [15, 19]. Although agile software development has become mainstream in industry [8, 9, 13, 14], the role of requirements engineering is still challenging in agile projects. Hard-core agilists believe that creating a requirements document at the beginning of a project is a waste of time because of the ever-changing terrain of the project. Hence, they neglect the usage of comprehensive requirement elicitation techniques [1, 6] and rely fundamentally on the use of validation and refactoring. However, several authors have recently pointed out that it is necessary to establish a balance between agility and intensive up-front activities in order to avoid high amount of refactoring, which would cause an escalation of development costs in later stages, possibly jeopardizing the whole project [7, 16]. Furthermore, practitioners recognize insufficient time for business analysis and problem examination as one of the serious problems, which

© Springer Nature Switzerland AG 2021
A. Przybyłek et al. (Eds.): LASD 2021, LNBIP 408, pp. 185–191, 2021.
https://doi.org/10.1007/978-3-030-67084-9_12

results in inadequate requirements. On top of that, requirements sometimes tend to be generated without a deeper reflection of their actual utility [3].

Although we agree that requirements are rarely to be found well-defined, and ready to be collected [10, 15, 18, 19], we believe that capturing initial requirements up front is a good idea. Furthermore, we share the view of Przybyłek [12] who claims that an information system should be developed with explicit consideration of the business processes it is supposed to support. Therefore, we have integrated his Business-Oriented approach to Requirements Elicitation (BORE) with Scrum. In BORE, system requirements are derived from business process models, which ensures that the system requirements meet real business needs and there are no superfluous requirements [12]. Further evidence indicates that process modelling can be used as a requirements documentation technique in agile projects [5].

Although BORE uses UML Activity Diagram to provide a seamless transition from business process models to use case models, we have decided to switch to BPMN (Business Process Model and Notation). The motivation for this change has been the growing importance of BPMN [2], which continues to dominate the process standards space [4]. BPMN is a standard for business process modelling that provides a graphical notation for specifying the end-to-end flow of a business process. The aim of this paper is to demonstrate that specifying initial requirements up front can be done in an agile and light way.

2   Case Study

This section reports on a Scrum project, named NiceAnt, in which we employed the aforementioned approach. The project was conducted in OKE Poland, which creates new multi-screen looks and functionalities including software for TV platforms and hybrid TV services. Before the NiceAnt project started, the project originator met with an analyst to discuss his vision and motivation. The motivation was based on a market analysis and feasibility study. The aim of the project was to create an advertisement website specialized in short-term exchange of employees between companies. That is, if Company A has a permanent employee who currently does not have any tasks assigned, he or she will be temporarily transferred to Company B, providing that specific requirements are met. Obviously, Company B is obliged to pay remuneration in the given period.

At the beginning of the project, the originator and analyst met several times to define the product backlog. Since the requirements were not out there in the project space waiting to be captured, they started by modelling the business processes. The created model helped them to pinpoint any possible errors in the workflow. Later on, the model was used to discover the real requirements represented by use cases, which served as a basis for further system development. In subsequent iterations, a few use cases were selected based on their priority for further analysis and implementation. The details of both models are presented in the succeeding subsections.

2.1   To-Be Process Model

Three business processes to be supported by the system are as follows: employee registration, employee sharing, and employee hiring. When Company A decides to share its employees, it has to register them into the system. An employee is added to the system by filling in information, including personal data, years of experience, education, type of employment (i.e. stationary or remote), position held, expected salary, place of residence, industry, skills, and foreign language proficiency. Next, the company can place this employee's offer immediately on the market. In addition, the following information will be provided, namely the period when the offer is valid and expected salary. Then, the new offer will be issued in the system and make available for searching by other companies.

The employee search process begins when the representative of another company uses the portal to search for an employee by specifying certain search criteria. Based on the query, the results are returned in a sorted way according to their relevance. If the user wants to see the full profile of a candidate, he or she has to pay service fee. After that, the user can start the negotiation process, which is further presented on Fig. 1.

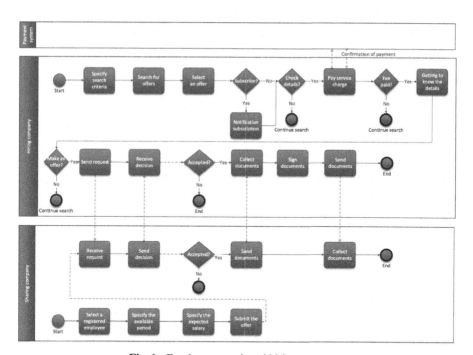

Fig. 1. Employee search and hiring processes

2.2 Use Case Model

The next step was to create a use case diagram (Fig. 2) based on the business process model. As a result, five actors and eighteen use cases were distinguished. Then, each element of the diagram was thoroughly described.

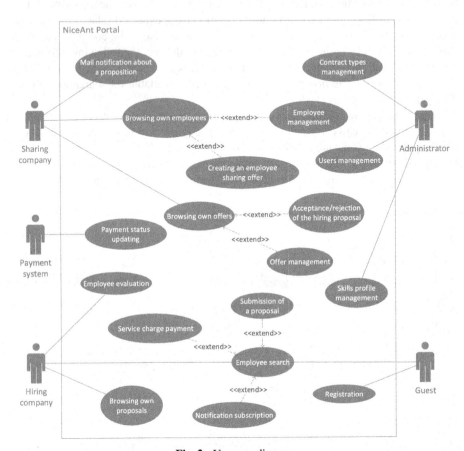

Fig. 2. Use case diagram

Description of the Actors:

- **Sharing company** – a company registered on portal NiceAnt, whose representatives provide sharing offers and accepts or rejects proposals from hiring companies.
- **Hiring company** – a company registered on portal NiceAnt, whose representatives search for temporary employees and negotiate contracts.
- **Administrator** – the one who manages system and its resources (e.g. offers, users, etc.).
- **Guest** – the one who can only search for available offers.
- **Payment system** – an external payment system for processing payment.

Description of the Use Cases:

- **Browsing own employees** – the sharing company browses the list of profiles created.
- **Employee management** – the sharing company manages employees' profiles (i.e., add, edit, and delete profiles).
- **Creating an employee sharing offer** – for the selected employee, a sharing offer including expected salary is created.
- **Browsing own offers** – the sharing company browses the list of offers it posted.
- **Acceptance/rejection of the hiring proposal** – the sharing company decides whether to accept the hiring company's proposal and then, proceed to negotiations.
- **Offer management** – the sharing company can update or remove out-of-date offers.
- **Registration** – a person who does not have an account can register in the system.
- **Employee search** – the hiring company or a guest can specify search criteria to find suitable candidates.
- **Notification subscription** – the hiring company subscribes to the notification for a specific offer or specifies an employee profile in which it is interested.
- **Submission of a proposal** – the hiring company responds to an interesting sharing offer.
- **Service charge payment** – the hiring company pays service charge in order to view an employee's detailed information and starts negotiations.
- **Employee evaluation** – the hiring company can assess an employee. The assessment will be visible to his or her employer.
- **Users management** – the administrator can update the users' data or remove the user accounts.
- **Skills profile management** – the administrator can add, remove, and edit the skills profiles.
- **Contract types management** – the administrator can add, remove, and edit the forms of employment that will be available in the system.
- **Payment status updating** – the payment system sends information confirming that the hiring company has paid the service charge.

3   Discussion

As a result of the analysis, the business process model was created for the three processes that were to be supported by the NiceAnt portal. The model presents the main activities performed in the system. A use case diagram was created based on the business process model and was further enriched with use cases that emerged as a result of discussions with the originator. Afterwards, the use case diagram was provided with descriptions of each actor and use case.

The agile team that developed the project expected quick results, as its originator was keen on creating a functional prototype as soon as possible. If the main objective was to reengineering processes of the existing system, the analyst could spend more time on scrutinizing the situation, which would produce richer models. Due to the urgency of the situation, a simpler approach had to be undertaken, that is, creating a business process and use case models in the most basic form. It turned out that a few meetings were enough to create both models and discuss them. Indeed, the discussion revealed new use cases. After the meetings, the analyst and the originator developed more systematic knowledge of the project. It shows that our approach brought added value to the project.

It should be remarked that the approach could be improved by inviting the development team to discuss the processes from their perspective. Nevertheless, the models were presented to the entire team, which also led to heated discussions.

4 Conclusion

The incorporation of requirements engineering practices is still an issue in agile development projects. Thereby, many information systems still do not fulfil the real needs of business. One of the problems detected in practice is the lack of overall understanding of the organization. As a remedy to this problem, we adopted BORE [12], which is a systematic approach to guide the process of generating use cases from the business process model. The approach allows system analysts to properly understand the organization and its environment in a participative way with all stakeholders. In this paper, we have demonstrated how the approach can be used to identify a set of requirements aligned with business objectives, in an agile project. The approach has turned out to be especially effective when requirements are not fully knowable up front and must be discovered. Moreover, the approach has been well accepted by the participant company. To improve the external validity of our findings, another case study is under way.

References

1. Eberlein, A., Leite, J.C.S.P.: Agile requirements definition: a view from requirements engineering. In: International Workshop on Time-Constrained Requirements Engineering, Essen (2002)
2. Gawin, B., Marcinkowski, B.: How close to reality is the "as-is" business process simulation model? Organizacija 48(3), 155–175 (2015). https://doi.org/10.1515/orga-2015-0013
3. Gawin, B., Marcinkowski, B.: Making IT global – what facility management brings to the table? Inf. Tech. Dev. 25(1), 151–169 (2019). https://doi.org/10.1080/02681102.2017.135 3943
4. Harmon, P., Wolf, C.: State of Business Process Management. A BPTrends Report (2016)
5. Jarzębowicz, A., Połocka, K.: Selecting requirements documentation techniques for software projects: a survey study. In: 2017 Federated Conference on Computer Science and Information Systems (FedCSIS), pp. 1189–1198, IEEE (2017).. https://doi.org/10.15439/2017f387
6. Jarzębowicz, A., Marciniak, P.: A survey on identifying and addressing business analysis problems. Found. Comput. Decis. Sci. 42(4), 315–337 (2017). https://doi.org/10.1515/fcds-2017-0016
7. Matkovic, P., Maric, M., Tumbas, P., Sakal, M.: Traditionalisation of agile processes: architectural aspects. Comput. Sci. Inf. Syst. 15(1), 79–109 (2018). https://doi.org/10.2298/CSI S160820038M
8. Mich, D., Ng, Y.Y.: Retrospective games in Intel Technology Poland. In: 15th Conference on Computer Science and Information Systems (FedCSIS), Sofia, Bulgaria (2020). https://doi.org/10.15439/2020f62

9. Ng, Y.Y., Skrodzki, J., Wawryk, M.: Playing the sprint retrospective: a replication study. In: Przybyłek, A., Morales-Trujillo, M.E. (eds.) LASD/MIDI -2019. LNBIP, vol. 376, pp. 133–141. Springer, Cham (2020). https://doi.org/10.1007/978-3-030-37534-8_7

10. Ossowska, K., Szewc, L., Weichbroth, P., Garnik, I., Sikorski, M.: Exploring an ontological approach for user requirements elicitation in the design of online virtual agents. In: Wrycza, S. (ed.) SIGSAND/PLAIS 2016. LNBIP, vol. 264, pp. 40–55. Springer, Cham (2016). https://doi.org/10.1007/978-3-319-46642-2_3

11. Przybyłek, A.: The integration of functional decomposition with UML notation in business process modelling. In: Magyar, G., Knapp, G., Wojtkowski, W., Wojtkowski, W.G., Zupančič, J. (eds.) Advances in Information Systems Development, vol 1, pp. 85–99 (2007). https://doi.org/10.1007/978-0-387-70761-7_8

12. Przybyłek, A.: A business-oriented approach to requirements gathering. In: 9th International Conference on Evaluation of Novel Approaches to Software Engineering (ENASE 2014), Lisbon (2014)

13. Przybyłek, A., Kotecka, D.: Making agile retrospectives more awesome. In: 2017 Federated Conference on Computer Science and Information Systems (FedCSIS 2017), Prague, Czech Republic, (2017). http://dx.doi.org/10.15439/2017F423

14. Przybyłek, A., Kowalski, W.: Utilizing online collaborative games to facilitate Agile Software Development. In: 2018 Federated Conference on Computer Science and Information Systems (FedCSIS 2018), Poznan, Poland (2018). https://doi.org/10.15439/2018f347

15. Przybyłek, A., Zakrzewski, M.: Adopting collaborative games into agile requirements engineering. In: 13th International Conference on Evaluation of Novel Approaches to Software Engineering (ENASE 2018), Funchal, Madeira, Portugal (2018)

16. Stal, M.: Refactoring Software Architectures. In: Agile Software Architecture, pp. 63–82. Morgan Kaufmann (2014). https://doi.org/10.1016/b978-0-12-407772-0.00003-4

17. Wawryk, M., Ng, Y.Y.: Playing the sprint retrospective. In: 14th Federated Conference on Computer Science and Information Systems, Leipzig, Germany (2019). https://doi.org/10.15439/2019f284

18. Weichbroth, P.: Delivering usability in IT products: empirical lessons from the field. Int. J. Software Eng. Knowl. Eng. 28(07), 1027–1045 (2018)

19. Zakrzewski, M., Kotecka, D., Ng, Y.Y., Przybyłek, A.: Adopting collaborative games into agile software development. In: Damiani, E., Spanoudakis, G., Maciaszek, L.A. (eds.) ENASE 2018. CCIS, vol. 1023, pp. 119–136. Springer, Cham (2019). https://doi.org/10.1007/978-3-030-22559-9_6

Keynote Paper

Pair Programming: An Empirical Investigation in an Agile Software Development Environment

Sanjay Misra[✉]

Covenant University, Ota, Nigeria
Sanjay.misra@covenantuniversoty.edu.ng

Abstract. Several experiments carried out on Pair Programming (PP) in a controlled environment by researchers and practitioners have been said to have a positive effect on software quality and time of delivery. Pair programming can be applied in all phases of software development. Although few empirical studies have shown the benefits of pair programming, not so much work has been done on maintainability of codes in a real agile environment. Therefore, in this work, we experimented using industry-based practitioners (working at an agile software development environment) to correct errors that were introduced deliberately into a set of python codes. Data was collected by recording the time to correct the mistakes. One hundred software practitioners were paired randomly and one hundred individual junior programmers to work on the same set of codes. Data obtained were analyzed, and we got very interesting results.

Keywords: Agile software development · Pair programming

1 Introduction

Several researchers have presented a comparison between existing software development models [1, 2]. In all those studies, one can find that a traditional software development approach, like the waterfall model, is not very practical in the situation of those projects where requirements are frequently changing. Furthermore, in conventional development, customers' requirements are fixed at the beginning of the project, and once the development has started, it is not possible to include new requirements. Therefore, more modern approaches to software development that will be dynamic enough to handle the present-day changing requirements were needed [3]. Hence, the emergence of Agile Software Development (ASD) method, which is a group of software development methods that gives room for software improvement because of its flexible response to change. ASD aims at delivering software products faster with high quality [4] and satisfying customer needs, adopting the principles of lean production to software development [5]. There are several benefits in ASD which overcome the shortcomings of the traditional software development as it focuses on keeping code simple, testing often and ensuring delivery of functional bits of the application as soon as they are ready. ASD is now a proved methodology for knowledge management and creativity [6].

© Springer Nature Switzerland AG 2021
A. Przybyłek et al. (Eds.): LASD 2021, LNBIP 408, pp. 195–199, 2021.
https://doi.org/10.1007/978-3-030-67084-9_13

Since the inception of agile software development (ASD) in 2001, a considerable number of software organizations have adopted and implemented various ASD methods for multiple tasks [7–9]. Amongst different ASD methodologies, Scrum [10] and Extreme programming (XP) are widely used agile methodologies. Researches are ongoing for increasing efficiency and productivity by combining various methods [11].

Pair programming is a practice used in XP - a popular agile software development methodology. In pair programming, two programmers are paired together to work on the same task (codes) using one system. They put ideas together on the same algorithm, same code and testing. One of the pair serves as the 'driver'; he types the input into the computer taking cognizance of the design or code while the other party acts as the 'navigator', who observes the work of the driver to pick errors and contributes objectively on how to rectify the mistakes [12]. Tactical and strategic defects detection and correction is the responsibility of the navigator [13]. Studies on pair programming have shown results in favour of quality and productivity. However, past findings show that pair programming is yet to be widely accepted practice, especially in industries.

Generally, it is believed and observed that there are numerous benefits of pair programming. According to [14], only about 22% of sampled programmers have practiced pair programming, but its application on a real project is just 3.5%. There are a number of benefits that can be derived from pair programming which include;

- Very few bugs are found in the software.
- Programmers end up producing very high-quality software.
- Pairs reason better than individuals.
- Partners have the benefit of exchanging ideas.
- Pairs can also gather experience from each other.
- Programmers' understanding of codes become better.

As good as these benefits are, there are salient but very important issues that if not looked into carefully and critically, may annul the numerous benefits of pair programming, making it unacceptable on real projects in software industries such as:

Conflict of ideas – not being able to reach a consensus on time, hence spending more time in discussion.

Inequality in the skill level of partners – one partner may not be as smart as the other; hence, he slows him down.

Personality incompatibility – these differences have a negative impact on productivity. Differences in lifestyle, temperament and value systems have bad effect on the process of software development in terms of product quality and time of delivery.

Distraction is another problem often encountered in pair programming as the navigator calls the attention of the driver to every little observation hence, slowing him down.

Programming and debugging style of pairs differ. There must be an agreement on what best style to adopt, and this is not quickly arrived at hence, having a serious impact on time.

By considering the above issues which reflect that apart from the several benefits of pair programming, there are several factors which hinder and reduce its efficiency and productivity. This point is the motivation of this present work. We have investigated how much pair programming is useful in tracing and rectifying the mistakes in real software development. Rigorous experimentations were carried out by combining various pairs and observing the impact of various types of pairing on identifying and rectifying the errors.

Previous studies and experiments in pair programming were carried out in an academic environment with students as the pair programmers. The involvement of students as pair programmers cannot be compared with professional pair programmers in a real environment as the productivity, quality and satisfaction of stakeholders will be completely different. Moreover, costs will be very difficult to estimate in a non-commercial academic environment.

The paper is structured in 3 sections. Section 2 presents the research methodology and conclusions on results. Conclusion drawn is presented in Sect. 3.

2 Research Methodology and Experimentation

The experiments were performed among software developers where data was acquired for the analysis. Deliberate errors were introduced into python codes and given to one hundred industry-based software practitioners to correct. These practitioners were located in North Central Nigeria (mostly in the capital city of Nigeria- Abuja). The time taken to correct the errors was recorded. The pairing was done randomly as shown in Table 1 with no regard to the expertise of the pairs in pair programming and then thereafter, only those with less than five years' experience (individual juniors) were examined.

Table 1. Grouping of software programmers

Grouping	Remark
Random pairs	Irrespective of their years of experience in pair programming
Individual junior	Less than five years of experience in agile programming

2.1 Results of Comparisons of Debugging Time for Various Pairs and Individual Programmers

Experimental results reflect the trend for the comparison in the time spent to debug 1 to 100 errors between Random pair and individual junior programmers (Fig. 1). The result shows (for 1–10 errors) that time spent by individual junior programmer was consistently higher than the random pair programmers although following a similar pattern. However, as soon as we move more than 10 bugs, the trend changed and debugging time for random programmer now became consistently higher than the individual junior programmers. This trend continued for up to 100 bugs [15]. The bugs were classified into groups of intervals of ten and averages of each group for different programmers were obtained to give the time taken per bug. Figure 1, shows the comparative average time of debugging for random pair and individual junior programmers. It was observed that between 1 to 50 bugs, individual junior requires higher debugging time than the random pair. The reverse is the case between 50 to 100 bugs.

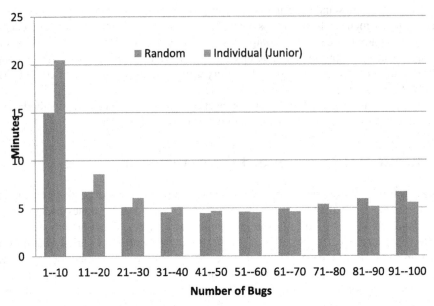

Fig. 1. Average time of debugging for grouped bugs

3   Conclusion

The result of the experiment on debugging different error complexities by random pairs and individual junior programmers revealed that the random pair spent the highest time of 21.88 min/bug on the average while individual junior spent 16.57 min/bug on the average [15]. The correlation analysis between the number of errors and time of debugging shows that the time spent increased significantly as the number of errors increased.

References

1. Misra, S., Omorodion, M., Fernández-Sanz, L., Pages, C.: A brief overview of software process models: benefits, limitations, and application in practice. In: Agile Estimation Techniques and Innovative Approaches to Software Process Improvement, pp. 258–271. IGI Global (2014). https://doi.org/10.4018/978-1-4666-5182-1
2. Patel, A., et al.: A comparative study of agile, component-based, aspect-oriented and mashup software development methods. Tehnicki Vjesnik **19**(1), 175–189 (2012)
3. Highsmith, J., Cockburn, A.: Agile software development. Bus. Innov. Comput. **34**(9), 120–127 (2001). https://doi.org/10.1109/2.947100

4. Poppendieck, T., Poppendieck, M.: Lean Software Development: An Agile Toolkit for Software Development Managers. Addison-Wesley, Boston (2003)
5. Cockburn, A.: Agile Software Development, p. 304. Addison Wesley Longman, Boston (2002). ISBN-10: 0201699699
6. de la Barra, C.L., Crawford, B., Soto, R., Misra, S., Monfroy, E.: Agile software development: it is about knowledge management and creativity. In: Murgante, B., et al. (eds.) ICCSA 2013. LNCS, vol. 7973, pp. 98–113. Springer, Heidelberg (2013). https://doi.org/10.1007/978-3-642-39646-5_8
7. Pham, Q.T., Nguyen, A.V., Misra, S.: Apply agile method for improving the efficiency of software development project at VNG company. In: Murgante, B., et al. (eds.) ICCSA 2013. LNCS, vol. 7972, pp. 427–442. Springer, Heidelberg (2013). https://doi.org/10.1007/978-3-642-39643-4_31
8. Zamudio, L., Aguilar, J.A., Tripp, C., Misra, S.: A requirements engineering techniques review in agile software development methods. In: Gervasi, O., et al. (eds.) ICCSA 2017. LNCS, vol. 10408, pp. 683–698. Springer, Cham (2017). https://doi.org/10.1007/978-3-319-62404-4_50
9. Rodriguez, G., Glessi, M., Teyseyre, A., Gonzalez, P., Misra, S.: Gamifying users' learning experience of Scrum In: Proceedings of ICTA 2020, CCIS. Springer, Heidelberg (2020)
10. Mundra, A., Misra, S., Dhawale, C.A.: Practical scrum-scrum team: way to produce successful and quality software. In: 2013 13th International Conference on Computational Science and Its Applications, pp. 119–123. IEEE, June 2013
11. Correia, A., Gonçalves, A., Misra, S.: Integrating the scrum framework and lean Six Sigma. In: Misra, S., et al. (eds.) ICCSA 2019. LNCS, vol. 11623, pp. 136–149. Springer, Cham (2019). https://doi.org/10.1007/978-3-030-24308-1_12
12. Beck, K.: Extreme Programming Explained: Embrace Change, Reading, Mass, pp. 10–70. Addison-Wesley, Boston (1999)
13. Bryant, S., Pablo R., Benedict, duB.: Pair programming and the mysterious role of the navigator. Int. J. Hum. Comput. Stud. **66**(7), 519–529 (2008). https://doi.org/10.1016/j.ijhcs.2007.03.005
14. Begel, A., Nagappan, N.: Pair programming: what's in it for me? In: Proceedings of the second AC-IEEE International Symposium on Empirical Software Engineering and Measurement. 9–10, October 2008, Kaiserlautern, Germany (2008)
15. Ajiboye, M.A.: Development of quality model metrics for agile software engineering, Ph.D. thesis, Federal University of Technology, Minna, Nigeria (2017)

Author Index

Printed in the United States
By Bookmasters